WORLD NAVAL AVIATION

WORLD NAVAL AVIATION

Paul Beaver

JANE'S DEFENCE DATA

First published in the United Kingdom in 1989 by
Jane's Information Group Limited
Sentinel House, 163 Brighton Road,
Coulsdon, Surrey CR3 2NX

ISBN 0 7106 0527 7

Distributed in the Philippines and the USA and its dependencies by
Jane's Information Group Inc,
1340 Braddock Place, Suite 300,
PO Box 1436, Alexandria,
VA 22313-2036

Typeset by Kerrypress, Luton, Bedfordshire.
Printed in the United Kingdom by Biddles Ltd,
Guildford, Surrey

Foreword

The importance of naval aviation has never been greater. Recent conflicts in the South Atlantic, Caribbean and the Gulf have seen the use of naval aviation in decisive combat situations. Although the main emphasis has been patrol, anti-submarine warfare, and search and rescue, the naval aircraft is now the system more capable than ever of projecting naval power.

There has been a spectacular increase in the number of maritime nations wishing to add missile-firing helicopters to their inventories during this decade. There are two reasons for this move into the anti-ship warfare role; firstly, the missile and sensor technology is now available for helicopters and secondly because of the success of anti-ship missiles in various small scale engagements.

In 1982, Exocet-carrying Argentine Navy Super Etendards sank HMS *Sheffield* and Mv *Atlantic Conveyor* in the South Atlantic. Later, towards the end of the Falklands (Malvinas) conflict, Lynx helicopters from UK Royal Navy destroyers, armed with Sea Skua, accounted for several Argentine patrol boats before that missile was even cleared for service. Five years later, USS *Stark* was severely damaged by two Exocet missiles launched from an Iraqi anti-ship tasked Mirage F1. Iraq's air force already operates helicopters in this role as well. Saudi Arabia and Chile have recently joined a long line of naval powers to acquire helicopters carrying missiles.

Strangely enough, the United States Navy has been slow to develop the idea of an anti-ship capable helicopter and only in the last few years has seriously looked at the advantages offered by the helicopter. Of course, the USN has been spoiled by having the world's largest and most capable fixed-wing strike force but it too has seen the flexibility being offered by the helicopter.

Whilst helicopters excel at the high hover time anti-submarine warfare sensor sweeps with dipping sonar or sonobuoys and have proved their ability for assault operations, they also carry out less obvious but equally vital roles such as vertical replenishment, airborne early warning and pilot training. Hardly a single new-build corvette, frigate or destroyer over 750 tonnes is not without its helicopter platform and even the USN's refurbished battleship squadron is capable of operating helicopters.

Carrier operations have also changed in the last 20 years with the advent of the short take-off and vertical landing strike fighter. The United Kingdom, USA, Spain and India now possess these aircraft, all derived from the Hawker Siddeley P1127 design of the 1960s. Italy will follow suit.

France is shadowing the USN's strike carrier concept with the building of one, possibly two nuclear-powered conventional aircraft carriers for the late 1990s to operate conventional fixed-wing aircraft.

The Soviet Navy has also taken to putting fixed-wing aircraft at sea with the Yak-36MP 'Forger' pioneering the way as a vertical take-off, short landing aircraft. But it too is looking for a conventional fixed-wing aircraft for its new composite-powered aircraft

Naval helicopters with operational anti-ship warfare capability.

Country	Helicopter	Missile
Brazil	Lynx	Sea Skua
Chile	Super Puma	Exocet*
France	Lynx	AS12
Germany (West)	Sea King	Sea Skua*
India	Sea King	Sea Eagle*
Indonesia	Super Puma	Exocet†
Iran	204 ASW	AS12†
Iraq	Super Frelon	Exocet
	Alouette III	AS12
Italy	Dauphin	AS15TT*
	AB 212ASW	Marte*
Korea (South)	Sea King	Marte*
	Lynx	Sea Skua*
Kuwait	Alouette III	AS12
Libya	Super Puma	Exocet
Pakistan	Super Frelon	Exocet
Peru	Sea King	Exocet
	Sea King	Exocet
Qatar (UAE)	AB 212 ASW	Marte*
Saudi Arabia	Sea King	Exocet
	Dauphin	AS15TT
Turkey	Super Puma	Exocet*
UK	AB 212 ASW	Sea Skua*
USA	Lynx	Sea Skua
USSR	Seahawk	Penguin*
	Ka-25	?
Venezuela	Ka-27	?
	AB 212ASW	Marte†
	Sea King	Exocet†

Notes: * indicates order yet to be fulfilled
† indicates confirmation still required

carrier currently fitting out in a Black Sea port. It seems that the Su-27 'Flanker' will be the baseline airframe rather than further developing the 'Forger', although in March 1988, the USN admitted the existence of a Yak-41 'Forger' follow-on. So the matter is still very unclear.

India is determined to increase its already sizeable maritime and naval air capability. A third aircraft carrier is now being planned and likely for the late 1990s, probably to replace *Vikrant*.

The world's two smaller aircraft carrier operators, Argentina and Brazil, continue to operate a mixture of fixed-wing strike and patrol helicopters. Depending on the task the air group's mix is changed. However, it can only be a matter of time before these operations become uneconomic and these countries confine their naval aviation to operating helicopters from escorts and land-based maritime patrol.

It seems that every manufacturer of airliners is promoting the use of regional, commuter, short-haul or medium-haul aircraft with high endurance to fulfil the needs of maritime patrol. It is a complex subject and the needs vary between armed and unarmed aircraft. In fact, it is often difficult to identify a nation's maritime patrol aircraft because they can have several roles, including transportation, training and special missions.

Author's Notes

World Naval Aviation is one of a number of Special Studies commissioned by Jane's Defence Data to fill key gaps in a specialised market. It is the first book which has been compiled to collate the inventory, organisational and operational facets of naval aviation. The term has been chosen to identify shipborne and land-based maritime aviation.

It has been relatively easy to identify shipborne aircraft and there are only a limited number of nations with fixed-wing inventories in this class. It is far more difficult to identify aircraft with a *primary* maritime role but which fly from land bases. Some nations appear to have no naval aviation at all even though they have coastlines. They could obviously use any aircraft to maritime patrol but without specialist adaptation they would not be viable for anything other than daylight, clear weather search and have therefore not been included in this study.

Acknowledgements

During the two years that this study has taken to produce, I have been assisted by a number of valuable sources. For part of the time, as naval editor of *Jane's Defence Weekly*, I had daily access to one of the world's best sources of current naval and aviation data and I am grateful to my colleague Peter Howard, managing editor of *JDW*, for continued access. Other Jane's Information Group team members have kindly assisted including Bridget Harney, the house-editor, Nick Cook, aviation editor of *JDW*, Derrick Ballington, chief librarian, David Steigman in Washington DC, and Captain Richard Sharpe, editor of *Jane's Fighting Ships*. Alec Sparks has completed the map artwork and Andrew Slade undertook much of the proof reading.

As usual with any author, I must take responsibility for the errors and omissions in this book.

Contents

Algeria

Organisations: Algerian Air Force; some transport aircraft appear to be registered to the Air Directorate of the Air Ministry for quasi-civilian operations. All military aircraft are operated by the air force.

Organisational structure: Algerian Air Force; appears to follow the Soviet regimental pattern but it suffers from inadequate training facilities under that system. There are four combat requirements and 13 squadrons. It is understood that substantial plans are underway to improve training and that France, the United Kingdom and West Germany have co-operated in flying and technical training programmes. Libya is also thought to co-operate with technical advisers.

Command structure: Thought to follow Soviet lines but shows French heritage. The Fokker F27 aircraft are assigned to naval operational control when on task.

Air-capable ships: None.

Shore bases: Alger; Bechar; Oran (Senia); Oran (Tafaraoui); Sidi bel Abbes; Setif; Constantie.

Embarked aircraft: None.

Shore-based aircraft: Beech Super King Air 200T (2); Canadair CL-215 (2); Fokker F27-400/600 (8).

Units: no data available.

Recent operations: Have been known to work with neighbouring states and in support of visiting Soviet warships.

Typical deployments: Offshore patrol and protection of EEZ with limited offensive capability. Close co-operation with the Polisario rebel group does not run to maritime operations, but it is understood that Morocco is considered to pose a potential threat.

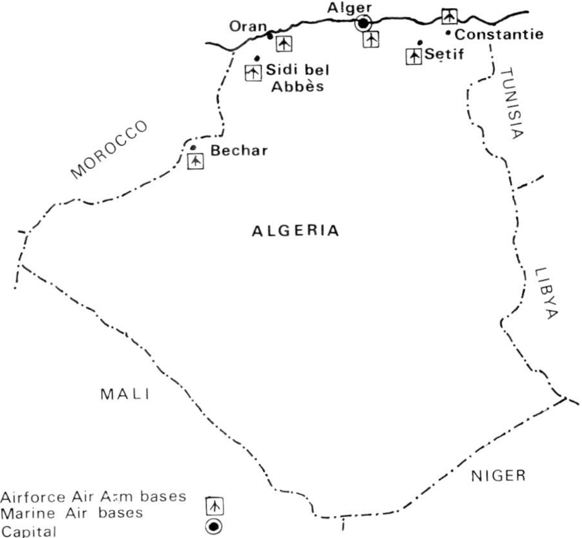

Weapon systems: All aircraft are thought to be unarmed, using weather and search radars only.

Personnel: 12 000 (total air force); 550 (coastguard).

Remarks: The Algerian Navy does not have a naval air arm and there are apparently no plans to form one.

Angola

Organisations: MPLA Air Force.

Organisational structure: Follows the Soviet regimental pattern with some sub-unit structures inherited from the Portuguese.

Command structure: Cuban and other Soviet Bloc advisers do not come under the MPLA's direct control. Some Soviet long-range reconnaissance aircraft have been observed operating from Luanda, in support of Angolan maritime patrols and for their own purposes.

Air-capable ships: None.

Shore bases: Ambriz; Ambrizete; Cabinda; Lobito; Luanada (Belas); Novo Redondo; Porto Amboim.

Embarked aircraft: None.

Shore-based aircraft: Aerospatiale SA 365F Dauphin (4); Embraer EMB-111 Bandeirante (2); Fokker F27 Mk600 Maritime (1).

Units: Squadrons and regiments appear to have no designations.

Recent operations: Although the air force is operating in support of ground forces in combat against the UNITA rebels in the eastern and southern provinces, and supports the SWAPO guerillas against Namibian and South African forces, it mainly concentrates on coastal patrol operations against insurgents.

Typical deployments: No overseas deployments have been made. Angolan maritime forces are supported by Soviet and Cuban aircraft.

Weapon systems: Embraer Bandeirante is armed with 127 mm rocket pods; Fokker F27 armed with up to 4 tonnes of depth bombs and torpedoes, although makes unknown at present.

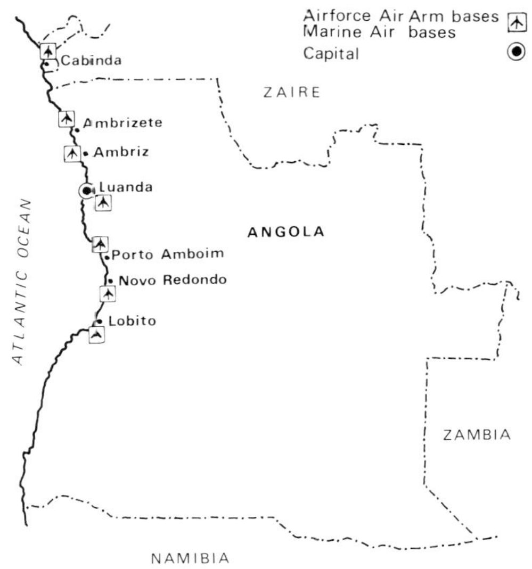

Personnel: 2000 (total air force). There are also at least 25 000 Cuban, 250 Soviet and 500 other Soviet Bloc advisers in the country, some of whom have been operating in support of the coastal patrols.

Remarks: Understood that Launda is used by Soviet long-range maritime patrol aircraft. 1986 delivery of Bandeirante only acknowledged in March 1987. Various French helicopters are on order or in the course of delivery; four Aerospatiale SA 365F Dauphin 2 could be used for maritime reconnaissance tasks but country's pre-occupation is with UNITA rebel and possible South African actions on land. Various helicopters could be made available for SAR, including IAR-316 Alouette III and IAR-330 Puma. The possible acquisition of missile-armed light aircraft could not be confirmed.

Angola operates a single F27 Maritime Mk600 for coastal surveillance duties, equipped with a Litton 360 degrees surveillance radar and bubble observation windows.

Argentina

Organisations: Naval Air Arm; Air Force; Prefectura Naval (Coastguard).

Organisational structure: Naval Air Arm; forms part of the Republic's Navy which is responsible to the Minister of Defence. The Coastguard is subordinate to the Navy and is closely modelled on the US Coast Guard.

Command structure: Naval Air Arm; divided into Wings and Squadrons; sub units thought to be flights, including those embarked. There are three Naval Areas which correspond to the coastal areas of the First, Second and Fifth Military Regions. Air Force: divided into brigades with two or more squadrons subordinate. Coastguard: aviation is sponsored by the Division de Aviacion.

Air-capable ships: Navy: 1 aircraft carrier (*25 de Mayo*); 2 Type 42 destroyers (doubtful operational ability); 4 MEKO 360 frigates: 4 MEKO 140 corvettes (2 building), 1 ice breaker (*Almirante Irizar*), 1 Antarctic supply ship (*Bahia Paraiso*); coastguard: 5 'Halcon' class patrol ships.

Shore bases: Naval Air Arm: Almirante Zar; Comandante Espora; El Plumerillo; Ezeiza; Gral Urquiza; Punta del Indio; Reconquista; Rio Gallegos; Trelew; Villa Reynolds.

Embarked aircraft: Naval Air Arm: Aerospatiale SA 319B Alouette III (7); Agusta ASH-3D/H Sea King (15); AMD-BA Super Etendard (14); Grumman S-2E Tracker (9); McDonnell Douglas A-4Q Skyhawk (24).

AMD-BA Super Etendards immediately after delivery in November 1981; the aircraft has recently returned to its carrier-based role. *(AMD-BA)*

Shore-based aircraft: Naval Air Arm: Aermacchi MB 326GB (4); Aermacchi MB 339A (5); Agusta A 109A Mk II (4); Embraer Xavante (12); British Aerospace 125 (1); Boeing CH-47C Chinook (2); Fokker F28 (3); Lockheed L-188E Electra (7); Air Force: Beech T-34C Mentor (11); Beechcraft Queen Air (5); Beechcraft King Air 200 (9); Lockheed C-130 Hercules (10); McDonnell Douglas A-4P Skyhawk (33); Coastguard: Aerospatiale SA 330 Puma (2); Aerospatiale AS 332B Super Puma (12); Bell 47G-3 (2); Douglas C-47 (2); Helibras HB 350B Esquilo (10)*; McDonnell Douglas 500M (6).

Units: Naval Air Arm: 1st Naval Air Wing (training); 2nd NAW (ASW); 3rd NAW (attack); 4th NAW (attack); 5th NAW (support); 6th NAW (reconnaissance); Naval Aviation School (Beech T-34C); Naval Aero-Photographic Squadron (Beech Queen Air/King 200); Naval Reconnaissance Squadron (Beech Queen Air); Naval ASW Squadron (Grumman S-2E); 1st Naval Helicopter Squadron

(Alouette III); 2nd Naval Helicopter Squadron (Sikorsky SH-3D); 1st Naval Attack Squadron (Aermacchi MB 326–339, Embraer Xavante); 2nd Naval Attack Squadron (AMD-BA Super Etendard); 3rd Naval Attack Squadron (McDonnell Douglas A-4Q); 1st Naval Logistic Support Sqn (Lockheed Electra); 2nd Naval Logistic Support Squadron (Fokker F28, BAe 125); Air Force: V Air Brigade (strike); I Escuadron Antartico; Coastguard: part of the Division de Aviacion.

Recent operations: Re-built following the Falklands conflict and expected to increase naval presence in Southern, South Atlantic and Pacific Oceans.

Typical deployments: Seasonal deployment to Antarctica (includes two Chinooks, ASH-3D and other helicopters) and continuous presence off Falkland Islands with continued interest in South Georgia.

Weapon systems: Naval Air Arm: primary anti-shipping missile is the Aerospatiale AM 39 Exocet whose operational ability was demonstrated in the Falklands conflict in 1982. Other weapons include conventional iron bombs, cannon and machine gun armament. Martin Pescador missile now operational on Xavante. For self-defence, Super Etendard can be armed with Matra Magic air-to-air missile and has a standard fit of two 30mm DEFA cannon. Magic can also be carried by Skyhawk and MB 339A aircraft.

Personnel: 3000 naval air arm; 20 000 total air force total: 9000 total coastguard.

Remarks: Two Navy Lynx were delivered in 1970s, one crashed in 1982 (not connected with Falklands conflict) and other sold to Brazil; continued deliveries of medium ASW helicopters from Italy and fixed-wing maritime patrol aircraft from Brazil. Argentina continues interest in AB 212ASW for embarked ASW and possible ASVW and in October 1987 four A 109A helicopters were delivered for maritime patrol and naval liaison tasks. Possible development of the Pucara aircraft for enhanced shipping strike role, fitted for embarkation aboard *25 de Mayo*. Doubt must be expressed about serviceability of remaining British equipment and five MB 339As thought to be in store. It is reported that the Super Etendards were primarily shore-based but now share the embarked strike role with the refurbished A-4Q Skyhawks. Five Electras were purchased in 1983 for maritime patrol conversion although one airframe was subsequently used for spares; recent airframe conversions from airliners gives a total flying inventory of seven at end of 1988.

Australia

Organisations: Royal Australian Navy (Fleet Air Arm); Royal Australian Air Force: National Aeronautical Safety Council.

Organisational structure: Fleet Air Arm: organised into three helicopter squadrons, the headquarters of naval flying is at Nowra, New South Wales; RAAF: control of most combat aircraft is through Operational Command which comprises 12 combat, transport and training squadrons with a headquarters at Glenbrook, New South Wales; NASC: a government funded organisation, centrally controlled from Canberra.

Command structure: Fleet Air Arm: follows the UK Fleet Air Arm with a Flag Officer responsible to the Chief of the Naval Staff: RAAF: the basic operational unit is the squadron which can be divided into flights for deployment; NASC: for SAR and other related tasks, helicopters are dispersed to RAAF stations.

Air-capable ships: RAN: 6 *Oliver Hazard Perry* class frigates, 1 amphibious warfare ship (*Tobruk*), 1 replenishment ship (*Success*), 1 destroyer tender (*Stalwart*), 1 training ship.

Shore bases: Fleet Air Arm: Nowra; RAAF (Maritime); NASC.

Embarked aircraft: Aerospatiale AS 350B Ecureuil (6); Sikorsky S-70B-2 Seahawk (16); Westland Sea King HAS 50/50A (7); Westland Wessex HAS 31B (3).

The four basic helicopters of the Royal Australian Navy: from the back, Westland Sea King HAS50, Westland Wessex HAS31B, Bell UH-1B Iroquois and Bell 206B JetRanger. *(RAN)*

Shipboard training for future naval helicopter pilots is provided in the Bell JetRanger helicopter, seen here approaching HMAS *Canberra's* flight deck. *(Bell)*

Shore-based aircraft: Fleet Air Arm: Bell UH-1B/C (3); Bell UM-1H (9); Bell 206B-1 JetRanger (3); BAe HS 748 (2); RAAF General Dynamics F-111C (15); Lockheed P-3C Update II Orion (20: McDonnell Douglas F-18B Hornet (57 total order); NASC: Bell 212 (10).

Units: Fleet Air Arm: HC-723 (AS 350B/UH-1/JetRanger); HS-816 (UH-1H Wessex); HS-817 (Sea

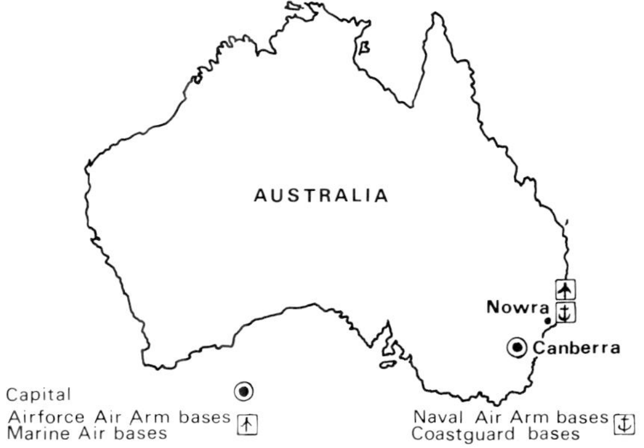

The General Dynamics F-111C strike aircraft provides the land-based anti-ship capability of the Royal Australian Air Force, armed with the Harpoon missile; weapons pictured here are laser-guided bombs. *(RAAF)*

King); RAAF: 1 Sqn (F-111C), 6 Sqn (F-111C), 10 Sqn (P-3C), 11 Sqn (P-3C); NASC: no system announced.

Recent operations: Following publication of Dibb Report in 1986, national defence interests have been paramount to Australia; in early 1980s the fixed-wing maritime aviation role was taken away from the RAN causing tremendous loss of morale and skilled personnel. In 1983, RAAF attempted to prove its ability to strike at maritime targets against a visiting British carrier group but was not fully successful. ANZUS treaty organisation now considerably strained through non-nuclear policy.

Typical deployments: Fleet Air Arm operates with deployed warships in Australian and neighbouring waters; occasional deployments to ASEAN nations and increased presence in Indian Ocean, RAAF operates long-range maritime reconnaissance, including South China Seas, Indian and Pacific Oceans, NASC provides increasing amount of SAR cover for RAAF.

Long-range maritime patrol is provided by the Lockheed P-3C Orion with a nose-mounted FLIR included in the equipment package. *(Paul Beaver)*

Weapon systems: Fleet Air Arm: helicopters are

Westland Sea King MK 50A is equipped with the Bendix ASQ-13B dipping sonar and the MEL ARI 5955 search radar; several of the type were modified for minesweeping trials 1987–88. *(Paul Beaver)*

Australia has begun to receive the Sikorsky S-70B-2 Seahawk helicopter.

capable of carrying the Honeywell Mk 44 (shallow water) and Mk 46 lightweight torpedoes, although the new weapon for the S-70B-2s has yet to be decided; British Aerospace Mk 11 Mod 0/1 depth bombs are also carried. RAAF: the P-3C Orions carry a full suite of ASW weapons including Mk 46 torpedoes and depth bombs, F-111 are iron bomb equipped but not capable of carrying nuclear weapons for political reasons. Australia has developed its own sonobuoy system for the P-3C and for possible fitting to the S-70B-2, which will also be equipped with the MEL Super Searcher radar. All F-18s capable of carrying AIM-9L Sidewinder and AIM-7 Sparrow. NASC: all aircraft unarmed.

Personnel: Fleet Air Arm: 16 500 (total); RAAF 22 000 (total): NASC: not known.

Remarks: The naval air force was drastically cut to fund the F-18 programme and all fixed-wing combat aviation was removed to the RAAF when the fixed-wing aircraft carrier, HMAS *Melbourne* was paid off in 1982. Plans to buy HMS *Invincible*, the light STOVL aircraft carrier from the UK Royal Navy were abandoned in 1982 and since then considerable effort has been made to fit all existing, new build and planned warships and auxiliaries with flight decks for naval helicopters. Two orders of total 16 Sikorsky S-70B-2 Seahawk made in 1986 for delivery in 1988/9; helicopters to be deployed aboard FFG-7 and new-build (Dibb Report recommended) frigates. Eight S-70B-2s to be assembled locally in Australia; complete weapon suite still to be announced. Fixed-wing F-111s are being standardised and brought up to date with EW systems parallel to those of the F-18 Hornet, including also the provision of Harpoon anti-ship missiles. It is understood that the P-3Cs will remain in service until 2005 following update programme which released P-3B aircraft to Portugal. In late 1988, nine Bell UH-1H helicopters were transferred from RAAF to RAN to supplement the three remaining UH-1Bs and the Wessex. Wessex HAS 31B will be phased out by end of 1989.

HS 748 electronic warfare aircraft. *(RAN)*

Bahrain

Organisations: Bahrain Defence Force Air Wing.

Organisational structure: The Air Wing operates as a homogeneous unit.

Command structure: The basic formation is the Wing, which is probably divided into Flights.

Air-capable ships: None.

Shore bases: Manama; Mubarraq.

Embarked aircraft: None.

Shore-based aircraft: MBB BO 105C (3).

Units: Air Wing.

Recent operations: None reported.

Typical deployments: Limited operations in support of the Bahrain Defence Force, State Police and Public Security.

Weapon systems: All helicopters are unarmed at present (see Remarks below).

Personnel: 200.

Remarks: It was reported at the 1988 Farnborough Air Show that the Bahrain Defence Force Air Wing has a requirement for two Aerospatiale SA 365F equipped with Thomson-CSF Agrion radar and the Aerospatiale AS 15TT anti-ship missiles. An order is expected by 1990.

Belgium

Organisations: Royal Belgian Air Force.

Organisational structure: The Royal Belgian Navy has operational control over the helicopters which are jointly owned/operated by the air force. The Royal Belgian Air Force is fully integrated into the NATO 2nd Allied Tactical Air Force command structure with about 140 combat aircraft in four wings.

Command structure: The basic formation is the Wing (Escadre) which is divided into two squadrons; the Wing reports to Tactical Air Force Command (fighter/strike aircraft) and the Transport Group (fixed-wing transport). Belgian Naval Aviation is confined to one squadron, the commanding officer is Major M Janssen; the staff officer (operations & training), Commandant L Dams; the SAR flight commander is Commandant M Clayes; the Naval flight commander is Lieutenant Commander E Verheyan.

Air-capable ships: 2 command ships (*Zinnia*, *Godetia*).

Shore bases: Antwerpen; Koksijde.

Embarked aircraft: Aerospatiale SA 319B Alouette III (3).

Shore-based aircraft: Westland Sea King Mk 48 (5).

Units: 40 Sqn (Alouette III/Sea King Mk 48).

Recent operations: No notable operations.

Typical deployments: Part of the multinational mine countermeasures force has deployed to the Gulf, the Alouette III helicopter embarked in *Zinnia* was ditched in July 1988 but recovered aboard and returned to operational service in 1989.

Weapon systems: Maritime aircraft are unarmed.

Personnel: 20 000 air force total.

Remarks: The single helicopter squadron is a joint air force/navy unit to support SAR and support ship operations. Maritime patrol by fixed-wing aircraft is not a Belgian NATO role and is carried out by the British and Dutch aircraft. The Gendarmerie/Rijkswacht operates three Aerospatiale SA 330C Puma helicopters which have an overwater capability but no operational role.

Westland Sea King Mk 48 in service with the Belgian Air Force for SAR/Combat Rescue duties; they are not equipped with weapon hardpoints. *(Paul Beaver)*

Three SA 319B Alouette III search and rescue helicopters are operated by the navy although jointly owned with the air force. A fourth Alouette III was lost in July 1988 when it ditched from BNS Zinnia in the Gulf. *(Paul Beaver)*

Belize

Organisations: Belize Defence Force.

Organisational structure: Chain of command small, leading to the Governor and Commander-in-Chief, with executive power held by the Prime Minister.

Command structure: Simple air wing follows British lines.

Air-capable ships: None.

Shore bases: Belize International Airport: Belmopan; Punta Gorda.

Embarked aircraft: None.

Shore-based aircraft: Pilatus Britten-Norman Defender (2).

Units: None

Recent operations: None

Typical deployments: Coastal patrol in support of British forces, with the main concentration of effort being devoted to EEZ protection and anti-drug smuggling operations.

Weapon systems: aircraft are armed with 7.62 mm machine guns and possibly rocket pods.

Personnel: 500 total (rising to 3000 in due course).

Remarks: Overall defence of this Commonwealth nation with the United Kingdom which stations Harrier fixed-wing strike aircraft, Puma light support and a flight of Gazelle liaison helicopters there to guarantee the nation's independence. Threat from Guatemala receding but increased risk of drug smuggling and underworld operations; no major expansion is forecast.

BN-2B Defender used by Belize Defence Force for patrol reconnaissance and liaison duties.

Bolivia

Organisations: Aviation of the Bolivian Naval Force.

Organisational structure: Unknown.

Command structure: Republic's president is commander-in-chief (Captain General) and the three armed forces are subordinate to the Supreme Council of National defence and its Minister. C-in-C of the Navy is the Chief of Staff.

Air-capable ships: None.

Shore bases: La Paz, Tarija.

Embarked aircraft: None.

Shore-based aircraft: Cessna U206G (1); Helibras HB 315B Gaviao (8); North American AT-6 (2).

Units: None

Recent operations: None.

Typical deployments: Patrol operations over and around Lake Titicaca are regularly undertaken; rivers also patrolled.

Weapon systems: The helicopters and the Cessna U206G (delivered in November 1980) are unarmed but the ex-Argentine AT-6 Texans have provision for light machine guns and unguided rockets.

Personnel: 4000 total navy.

Remarks: Bolivia is a land-locked country but a small naval force is maintained to patrol the lakes and rivers, especially along the Peruvian border. It is understood that Bolivian Carabineros MBB BO 105C and Cessna 421B aircraft would be made available to the naval forces if required.

Brazil

Organisations: Brazilian Air Force: Brazilian Navy.

Organisational structure: State president is Supreme Commander with three major services subordinate through the Minister of National Defence via their own General Staff. There are seven Navy and six Regional Commands based on geographical areas. The Brazilian Air Force operates about 150 combat aircraft and a further 50 are under naval control.

EMBRAER EMB-111 Bandeirante maritime patrol aircraft is designated P-95 in Brazilian Air Force service. The main sensor is the AN/APS-128 search radar and primary weapon is the 127 mm HVAR rocket.

Command structure: Within the air force, the subordinate Comando Costeiro (COMCOS) controls naval patrol aviation with the 7^{C} Grupo de Aviacao operating two squadrons (1^{0} & 2^{2} Esq) and the 10^{0} Grupo (SAR) with two squadrons (2^{0} & 3^{0} Esq). Naval aviation is based on the squadron but includes detached flights at sea, bearing a close relation to the US Navy.

Air-capable ships: 1 aircraft carrier (*Minas Gerais*), 2 *Gearing* class destroyers, 4 *Allen M Sumner* (FRAM II) class destroyers, 4 ocean patrol frigates (building), 6 Vosper Thornycroft MK 10 frigates, 1 Antarctic exploration ship, 2 *Pedro Teixeira* class river patrol craft, 2 *Sirius* class hydrographic ships, 1 cadet training ship, 1 lighthouse tender.

Shore bases: Air Force: Campo Grande (2^{0} Esq, 10^{0} Grupo); Florianapolis (2^{0} Esq, 7^{0} Grupo, 3^{0} Esq, 10^{0}

Bell JetRanger III used for pilot training by the Brazilian Navy and equipped with emergency flotation gear on the skids. *(Bell)*

Westland Lynx Mk 21 helicopters equip Vickers Mk 10 frigates and most are now Sea Skua-capable. This is the first delivery still wearing UK class B registration. *(Westland)*

Grupo); Recife (1º Esq, 7º Grupo); Santa Cruz (2º Esq, Grupo de Aviacao Embarcada; navy: Sao Pedro de Aldeia (HQ); Salvador.

Embarked aircraft: Navy: Aerospatiale AS 350B Ecureuil (9); Aerospatiale AS 332F Super Puma (10); Aerospatiale AS 355F Ecureuil 2 (20); Agusta-Sikorsky ASH-3D Sea King (8); Bell TH-57B SeaRanger (15); Westland Lynx MK 21/23 (8); Westland Wasp HAS 1 (3); Air Force: Grumman S-2A Tracker (7); Grumman S-2E Tracker (6).

Shore-based aircraft: Navy: Agusta ASH-3H Sea King (4); Bell TH-57A SeaRanger (12); Bell UH-1H (12); Helibras HB 350B Esquilo (20); Sikorsky SH-3A (4); Westland Whirlwind HAR 9 (3); Air Force: Embraer EMB-110B Bandeirulha (14); Embraer EMB-111 Bandeirante (10), (10)*; Lockheed RC-130E Hercules (3).

Helibras HB 350B Esquilo carrier-borne SAR helicopter aboard *Minas Gerais* during flight deck firefighting drills *(Paul Beaver collection)*

Units: Air Force: Grupo de Aviacao Embarcada (embarked fixed-wing); 7º Grupo (maritime patrol) 1º Esq (Bandeirante), 2º (Bandeirulhas); 10º Grupo (SAR), 2º Esq (Bandeirante), 3º Esq (UH-1H, SH-1H); Navy: 1º Attack Helicopter Sqn (Sikorsky SH-3A); 1º General Purpose Helicopter Sqn (Whirlwind, Wasp, Lynx); 2º General Purpose Helicopter Sqn (Super Puma); 1º Helicopter Training Squadron (TH-57).

Recent operations: None.

Typical deployments: Long range maritime patrol flights are carried out into the South Atlantic and around the long coastline of the country; regular naval exercises with US and South American navies are carried out. The newly delivered Super Puma helicopters are used to support Brazilian Marines in assault and other amphibious operations.

Weapon systems: The standard lightweight aerial ASW torpedo is the Honeywell Mk46 Mod 2 although there are aerial depth bombs and rockets carried by the S-2Es. The recently delivered British Aerospace Sea Skua helicopter-launched anti-shipping missile gives the frigates an over-the-horizon capability. Fleet air defence would be presumably provided by air force combat aircraft but no details are available.

Personnel: 35 000 air force total; 46 000 navy total.

Remarks: Following the re-introduction of naval aviation in 1960, it was agreed that the air force would retain all fixed-wing aircraft, including those embarked (Grupo de Aviacao Embarcada) but that the navy would operate the helicopters. Re-equipment is proceeding within the constraints of the national debt. A further ten Bandeirante aircraft were ordered in 1987 for delivery commencing in 1989, part of this order is to replace the RC-130E Hercules; equipment updates with improved avionics will follow for all

Westland Wasps are armed with the Avibras SBAT-70 rocket pod for anti-ship and close support tasks. Brazilian Navy Wasps are not armed with guided weapons. *(Avibras)*

remaining EMB-111 aircraft including updated processing for the MEL Sea Searcher radar. The Lynx helicopters have recently received Sea Skua anti-shipping missiles in a deal which is believed to be linked with adoption of the Tucano training aircraft by the UK Royal Air Force. A re-engining programme is underway for the Lynx and a new 360 degrees radar is planned. Further orders for light shipborne helicopters can be expected with Westland Super Navy Lynx and the Aerospatiale SA 365F Dauphin in competition. Confusion still continues over the exact number of model 350 single-engined helicopters, with differing official returns from the Brazilian Navy, the French manufacturer Aerospatiale and the Brazilian joint-venture company Helibras. SH-3A Sea Kings to be replaced by late model Agusta-built variants.

Bulgaria

Organisations: Bulgarian Navy.

Organisational structure: Follows Soviet lines.

Command structure: Follows Soviet lines with the basic unit being the Regiment which is usually divided into three squadrons.

Air-capable ships: None.

Shore bases: Balchik; Burgas; Varna (also Naval HQ).

Embarked aircraft: None.

Shore-based aircraft: Mil Mi-2 'Hoplite' (2); Mil Mi-4 'Hound' (6) (transport only), Mil Mi-14 'Haze A' (12).

Units: One squadron flies Mi-2 and Mi-4; Mi-14 in newly formed unit but no further details are known.

Recent operations: None reported.

Typical deployments: Coastal surveillance and offshore patrol in support of own naval craft and Soviet Black Sea Fleet; also supports Danube River Flotilla.

Weapon systems: Mi-2 and Mi-4 unarmed; Mi-14 operates with the standard Soviet torpedoes, depth bombs and mines; no nuclear warheads come under the command of Bulgarian naval forces.

Personnel: 10 000 total naval strength.

Remarks: Until the recent delivery of the MiG-23 for air defence duties, the Bulgarian armed forces were the smallest and least well equipped force in the Warsaw Pact. Several Antonov An-14 light transports are operated for Coast Guard duties by the Ministry of Fisheries.

Burma

Organisations: Union of Burma Air Force.

Organisational structure: Little data is available but it is thought a total of 50 aircraft are operated.

Command structure: Little data available but the squadron seems to be the basic unit.

Air-capable ships: Hydrographic ship (*Thu Tay Thi*).

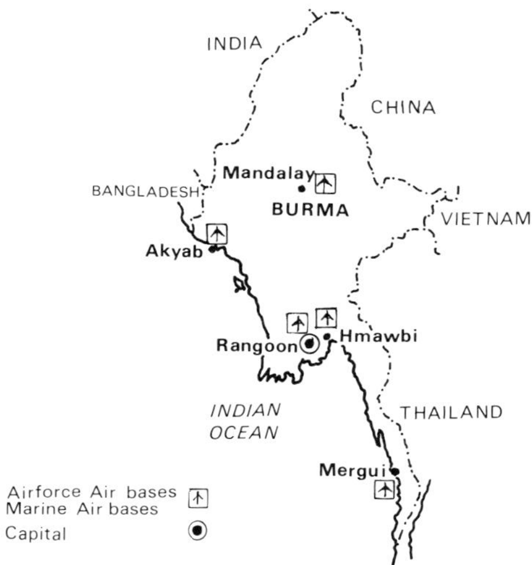

Shore bases: Akyab, Hmawbi, Mandalay, Mergui, Rangoon (Mingaladon).

Embarked aircraft: Aerospatiale SA 316B Alouette III (10), Kawasaki-Bell 47G-3 (10).

Shore-based aircraft Fokker F27M (3); Kaman HH-43B Huskie (6); Kawasaki KV-107-I (2).

Units: Information not available.

Recent operations: None.

Typical deployments: Coastal patrol in support of national EEZ, with the main concentration of effort towards anti-drug smuggling operations. One helicopter is regularly deployed to the hydrographic ship and other rotary-wing craft are used for search & rescue.

Weapon systems: Helicopters are armed with 7.62 mm machine guns and possibly rocket pods.

Personnel: 7000 (total).

Remarks: Some confusion reigns about the exact designation of the F27 aircraft as some sources give the aircraft as second-hand Fairchild FH-227 Friendships. Despite a long coastline but because of severe funding restrictions, the Union of Burma Air Force does not have dedicated maritime patrol aircraft.

Cameroon

Organisations: Cameroon Air force.

Organisational structure: The air force is responsible for all military aircraft operations.

Command structure: The basic operational unit is the Squadron, based exclusively on French Air Force lines.

Air-capable ships: None.

Shore bases: Batouri, Douala, Garoua, Yaounde.

Embarked aircraft: None.

Shore-based aircraft: Dornier Do 128-6MPA (3).

Units: All maritime patrol aircraft are operated by one squadron with an unknown designation.

Recent operations: None.

Typical deployments: Aircraft operate for anti-smuggler operations and the protection of the EEZ.

Weapon systems: The aircraft are unarmed.

Personnel: 350 total.

Remarks: Although the air force transport fleet (DHC Buffalo/C-130H Hercules) could operate for maritime patrol, the nation has the dedicated service of three Do 128 twin-engined aircraft, although recent reports say that only two are kept airworthy at any one time for budgetary reasons.

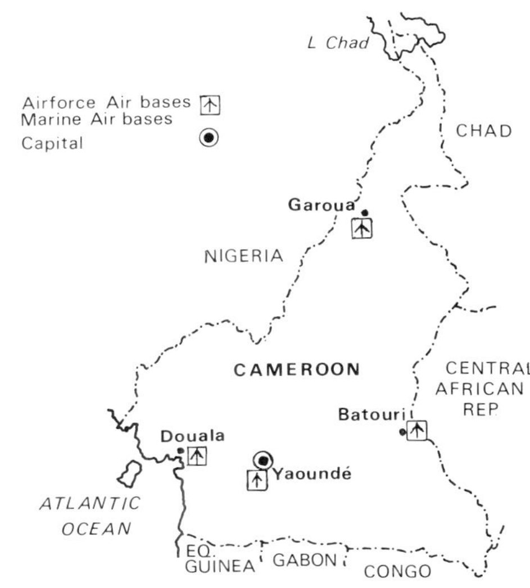

Canada

Organisations: Canadian Armed Forces.

Organisational structure: The Air Command administers the aircraft of the CAF but operational control rests with Maritime Command (HQ Halifax, Nova Scotia). The Maritime Air Group is headquartered at Halifax where it administers the fixed-wing and rotary-wing aviation for both Oceans. The Air Transport Group administers the 'air force' Search & Rescue operations based at Summerside, Prince Edward Island and Trenton, Ontario. The Air Reserve Group controls a Tracker-equipped reserve squadron at Summerside.

Command structure: The basic operational unit is the Squadron, organised more on the US Navy line than that of the UK Royal Navy, but retaining the British allocated squadron numbers from the Second World War. The Commander, Air Command in 1988 was Lieutenant General L A Ashley CMM CD and the Commander, Maritime Air Group, Brigadier C M A Curleigh DMM CD.

Air-capable ships: 4 *Iroquois* class destroyers, 6 *Halifax* class frigates (building), 2 *Annapolis* class frigates, 6 *St Laurent* class frigates, 2 replenishment oilers, 1 underway replenishment ship.

Shore bases: Comox (maritime patrol/helicopters); Greenwood (maritime patrol); Shearwater

(helicopter/utility); Summerside (maritime patrol); transport and rescue: Summerside (helicopters); Trenton (helicopters).

Embarked aircraft: Sikorsky CH-124A Sea King (34).

Shore-based aircraft: Bell CH-135 Twin Huey (45); Boeing CH-113A Voyageur/Labrador (14); de Havilland Canada CC-115 Buffalo (14); Grumman

Long-range SAR is flown by the Boeing CH-113A Voyageur helicopter (this example is from 442 Squadron, Canadian Armed Forces). An update and modification programme was completed in 1983. *(Boeing)*

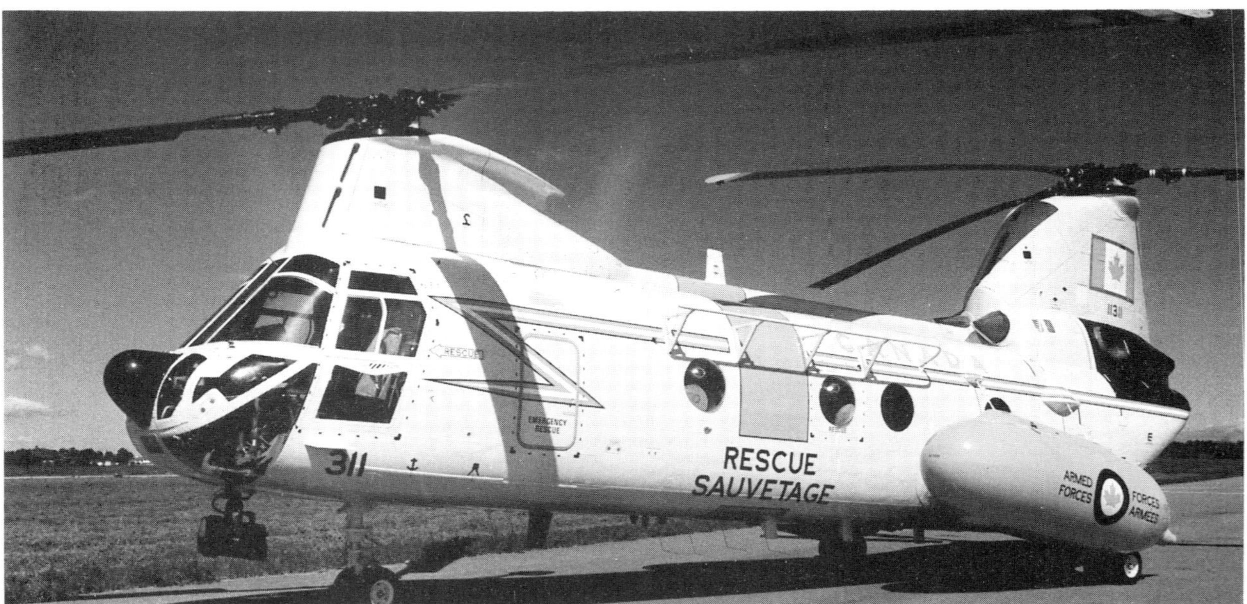

The Lockheed CP-140 Aurora is the standard long-range maritime patrol aircraft, operating over the Atlantic, Arctic and Pacific Oceans. *(Canadian Forces)*

CP-121 Tracker (15); Lockheed CP-140 Aurora (18); Lockheed CT-133 Silver Star (15).

Units: Maritime Air Group: 404 Sqn (CP-140); 405 Sqn (CP-140); 406 Sqn (CH-124A); 407 Sqn (CP-140); 415 Sqn (CP-140); 423 Sqn (CH-124A); 443 Sqn (CH-124A): VU-32 (CH-135/CT-133); 880 Sqn (CP0-121); VU-33 (CH-135/CT-133); Air Transport Group: 413 Sqn (CH-113A/CC-115); 424 Sqn (CH-113A/CC-115); Reserve Air Group: 420 Sqn (CP-121).

Recent operations: The Canadian Armed Forces has recently increased its vigil on the Arctic Ocean where there is considered to be an increasing Soviet military threat and an increasing US commercial interest.

Typical deployments: From early 1987, a flight of CH-124A has been deployed to the Pacific coast with two DDHs as part of a strengthening of the Western maritime presence. Canadian maritime aircraft regularly deploy to Western Atlantic bases and to Europe. Annually one or two CP-140 Auroras take part in the Fincastle fixed-wing ASW trophy.

Weapon systems: Anti-submarine warfare: some Canadian aircraft are known to be capable of carrying a nuclear depth bomb. Conventional weapons include the Honeywell Mk 44 and Mk 46 lightweight torpedoes; depth bombs, sensors: a full range of disposable sonobuoys and the CH-124A is equipped with the Bendix ASQ-13 dipping sonar.

Personnel: Air Command 23 000 (total); Maritime Command 8500 (total).

Remarks: There is a two-Ocean naval policy with increasing emphasis being given to the Pacific and in the next decade it is expected that there will be a third Ocean policy with the Arctic Ocean playing an increasing role in Canadian defence planning.

Coastal and EEZ patrols are mounted by the Grumman CP-121 Tracker patrol aircraft, flying from CFBs Comox and Shearwater. The Tracker will undergo a modernisation programme to equip it for a further 10 years service. *(Canadian Forces)*

The primary ASW system of the 'Tribal' class destroyers is the Skorsky CH-124A Sea King helicopter, seen here in the post-1987 low visibility colour scheme with two 'Tribals' in the background. *(Canadian Forces)*

Future programmes

Aurora Improvement Programme: Maritime Command hopes to improve the acoustic and other systems carried by the CP-140 Aurora long-range maritime patrol aircraft, including the installation of the CAE Electronics advanced integrated magnetic anomaly detector.

Helicopter Acoustic Processing System: The advanced acoustic processing system for the NSA developed by Computing Devices of Canada.

Helicopter Integrated Navigation System: Being developed by Honeywell Canada for the NSA programme for tactical navigation at sea.

Helicopter Integrated Processing and Display System: Linked to the HAPS for the NSA helicopter under a joint development by Computing Devices, Canadian Marconi and Litton Systems. It provides the ASW action crew with displayed information on the acoustic and tactical picture.

New Shipborne Aircraft: Maritime Command intends to replace the existing fleet of CH-124A Sea King helicopters with the Anglo-Italian EN 101 Merlin helicopter following definition phase contract by the Department of National Defence. Between 30 and 50 helicopters will be required under the aegis of the NSA project, mainly for the 'Tribal' class improvement programme and the new-build *Halifax* class Canadian Patrol Frigates. The Merlin in Canadian service will carry four lightweight torpedoes or two McDonnell Douglas Harpoon anti-ship missiles.

Tracker Improvement Programme: The CP-121 Tracker will be modernised for its continued coastal and exclusive economic zone patrol duties in the 1990s. Plans include the replacement of the existing piston engine with a turboprop and more advanced sensor gear including FLIR and the Computing Devices AN/UYS-503 acoustic processor.

Chile

Organisations: Chilean Naval Air Arm.

Organisational structure: President is Commander-in-Chief, but in peacetime the Minister for National Defence has the day-to-day responsibility of running the armed forces. There are three Naval Zones. The naval air arm is styled Naval Aviation Command in English.

Command structure: Basic operational unit for naval aviation is the squadron, based on the UK Royal Navy pattern, including the deployment of helicopter flights to warships. In 1988, The Chilean Navy won control of all over water military aircraft. The commander of the Chilean Naval Aviation Command in 1988 was Rear Admiral Claudio Aguayo and his chief of staff Captain Fernando Sarabia.

Air-capable ships: 4 'County' class destroyers, 2 *Allen M Sumner* (FRAM II) class destroyers, 2 *Leander* class frigates, 1 *Tide* class replenishment ship, 1 Antarctic patrol ship, 3 *Batral* class landing ships.

Shore bases: El Belloto (Valparaiso); Punta Arenas; Talcahuano. Pues to Williams Concón Sur.

Embarked aircraft: Aerospatiale SA 319B Alouette III (8); Bell 206B JetRanger (3).

Shore-based aircraft: CASA 212 Aviocar (4); AMD-BA Falcon 20/200 (1)*; Embraer EMB-110 CN (3); Embraer EMB-111 AN Bandeirante (6); Enaer/CASA A-36 Halcon (9) (11)*; Pilatus PC-7 Turbo-Trainer (10); Piper PA-31 Navajo (1).

Units: VC-1 (CASA 212, Bandeirante, Navajo); HS-1 (Alouette III, Bell 206B); VP-3 (Bandeirante); VT-4 (PC-7).

Recent operations: None.

Typical deployments: Replenishment ships can act as helicopter carriers. Chile continues to monitor Argentine naval operations and is increasingly concerned about illegal fishing off its long coastline.

Until the arrival of the new Dauphin helicopters, the Alouette III remains the primary shipborne helicopter. This example from squadron HS-1 is equipped with nose-mounted search radar and carries an FN Herstal 7.62 mm twin machine pod.

Aerospatiale commenced delivery of missile-armed AS 332F1 Super Puma helicopters in 1988. This is a trials aircraft shown armed with two AM 39 Exocet missiles.

Weapon systems: Alouette III armed with Mk 44 Mod O ASW torpedo (possibly locally modified Honeywell Mk 46 Mod 1 in service); A-36 reported to be fitted for BAe Sea Eagle anti-shipping missile; EMB-111 and PC-7 use 127 mm rockets for ASV.

Personnel: Naval Air Arm: 500.

Remarks: ASV operations will be extended with acquisition of ASM, such as BAe Sea Eagle for the Halcon and the procurement of new naval helicopters. Chilean Navy selected Lynx (small ship) and Sea King (DLG) but Westland were unable to deliver Lynx in time and US government banned Sea King sale, so French helicopters have been ordered. The SA 365F Dauphin will replace the Alouette III and the new anti-shipping type is the NAS 332F Super Puma, manufactured under licence in Indonesia. Super Pumas are primarily for ASW but will be Exocet-capable; the Dauphins have no anti-shipping capability at present. Four additional *Leander* frigates to be acquired.

Maritime patrol operations are carried out by EMBRAER Bandeirante aircraft equipped with Litton AN/APS-128 radar and a variety of lights. The aircraft is capable of deploying sonobuoys, flares and other disposables. Chile's Bandeirantes, delivered in 1978/79, have full de-icing and passive electronic surveillance measures equipment fitted.

Communications and liaison transport tasks are undertaken by EMB-110 aircraft, operated by squadron VC-1.

Pilatus PC-7 trainer serves with squadron VT-1 for a range of tasks including light armed reconnaissance.

China

Organisations: Navy of the People's Liberation Armed Forces.

Organisational structure: The exact structure of the naval element within the PLA is difficult to appreciate but it is known that the naval air arm is under the subordinate control of the navy which itself is subordinate to the Central Military Committee of the Chinese Communist Party. According to an official response, the force is called in English, the Chinese People's Naval Aviation Arm (CPNAA) and was founded as such in 1952. In 1988 the commander was Vice Admiral Li Jing, the Deputy Commander is Huang Decheng, the political commissar is Xing Yongning and Shan Dade is his deputy.

Command structure: It is understood that CPNAA units are organised into Air Divisions of three Regiments. Each regiment consists of three Companies or Groups, each with three Squadrons. The Chinese People's naval forces are divided into three fleets: East Seas, North Seas and South Seas.

Air-capable ships: 6 *Modified Jianghu* class frigate, 2 *Yuan Wang* class satellite/missile tracking ships, 1 *Xiang Yang Hong* class experimental ship, 3 *Da Jiang* class submarine support ships, 3 *Fuqing* class replenishment oilers (AOR).

Shore bases: Fuzhou, Guangzhou, Haikou, Hangzhou, Jinan, Luda, Ningpo, Qingdao, Shanghai, Tanjin, Xiamen, Yantai, Zhanjiang.

Embarked aircraft: Aerospatiale SA 321G Super Frelon (13); Harbin Z-8 (Super Frelon) (2); Harbin Z-9A (Dauphin 2) (10) (40)*.

Shore-based aircraft: Beriev Be-6 'Madge' (12) Hanzhong Y-8 (1), Harbin H-5 'Beagle' (80), Harbin Z-5 'Hound' (40), Nanchang JZ-5 (40), Nachang JZ-6 (30), Nanchang Q-5 'Fantan' (100), Shenyang J-5 'Fresco' (150), Shenyang J-6 'Farmer' (300), Shenyang JJ-2 'Mongul' (20), Shenyang JJ-5/JJ-6 (150), Tupolev Tu-16/Xian H-6 'Badger' (35), Xian J-7 'Fishbed' (200). Some J-8B 'Finback' aircraft were seen at Yantai in 1986, probably as a trials unit.

Units: Traditionally, the East Seas Fleet has two Air Divisions (1st and 6th Air Division), plus the elite, Hainan-based Independent Air Regiment. This is actually Division-size and could now be re-designated the 5th Air Division. North Sea Fleets has two Air Divisions (2nd and 3rd). South Seas Fleet has the 4th Air Division in support. According to local sources this gives 15 regiments (45 companies) of air defence and strike aircraft, at least three regiments (15 companies) of helicopters and two

One of the first Harbin Z-9A shipborne helicopters from the Chinese production line established by Aerospatiale. It is reported that China is seeking the missile-armed version of the Dauphin to equip its new frigates for anti-shipping duties. *(CATIC)*

regiments (four companies) of up to 80 transport aircraft. Since 1975, there has been a massive expansion programme with emphasis on the South Seas Fleet units and it is now understood that there were 10 Air Divisions (30 Regiments) with a further five Independent Air Regiments in 1985.

Recent operations: Considerable efforts have been made in the past five years to increase the operational efficiency of the embarked aviation force, especially with the development of helicopter-carrying frigates. Previous helicopter operations had been confined to the auxiliary ships. Recent Vietnamese naval operations have been monitored by the CPNAA from Hainan and other South Seas Fleet locations.

Typical deployments: The Chinese Navy is not a blue water force as yet and so most of the regular training and deterrent patrols are carried out along the coast, with special emphasis on the southern area (adjacent to Vietnam) and the north of the Yellow Sea (close to the Soviet Pacific Fleet).

Weapon systems: Shenyang H-5 'Beagle' strike aircraft are equipped with air-launched torpedoes and depth bombs; fighter aircraft, like the Shenyang J-5 and J-6 are armed with cannon, rocket pods and guided weapons of limited ability. Xian H-6 has been pictured since 1984 carrying the C-601 air-launched cruise anti-shipping missile. Helicopters appear to be unarmed, although official statements credit the Harbin Z-8 and Z-9 with short and medium ASW capability respectively, and this must include lightweight torpedoes and depth bombs. Some of the SA321G Super Frelons are shore-based for SSBN escort duties. Three Super Frelons have been fitted with the Thomson Sintra HS-12 dipping sonar and HS 312S sonic processor for this escort role.

Personnel: 30 000 (naval air arm).

Remarks: Future plans include the expansion of the embarked helicopter force, more embarked troop-carrying helicopters and the possible development of an embarked fixed-wing force using an aircraft carrier. This latter project has not been mentioned in official literature for some years and sources indicate that maritime patrol aircraft will be acquired in preference to carrier-borne aircraft. Weapon development continues, particularly with anti-shipping missiles. New destroyer and frigate programmes may require a better dedicated ASW/ASV helicopter than the Z-9A and reports indicate that the SA 365F Dauphin is under consideration for small ship ASW role.

The first picture of the HAMC SW-5 amphibian taking to the water. Note the dorsal gun position and the nose-mounted search radar. Development of this aircraft is now proceeding. *(CATIC)*

Colombia

CARIBBEAN SEA
VENEZUELA
PACIFIC OCEAN
Bogota
Buenaventura
COLOMBIA
BRAZIL
ECUADOR
PERU
Naval Air Arm bases
Coastguard bases
Capital

Organisations: Colombian Naval Air Arm; Colombian Air Force.

Organisational structure: Head of the armed forces is the highest ranking army general, although naval command comes through the naval C-in-C.

Command structure: The Colombian naval air arm is a new development, reconstructed after several years of total air force domination of the maritime task. At present, there is a small helicopter unit only. The air force provides four strike aircraft for naval duties but under air force operational control.

Air-capable ships: 4 FS 1500 frigates.

Shore bases: Buenaventura.

Embarked aircraft: navy: MBB BO 105CB (4), (4)*.

Shore-based aircraft: air force: Cessna A-37B Dragonfly (4).

Units: Helicopter unit.

Recent operations: None.

Typical deployments: The helicopters deploy to their parent frigates on an required basis, including taking part in multi-national naval exercises.

Weapon systems: The air force A-37Bs carry 7.62 mm machine guns and have eight pylons for bombs or rockets up to 2.5 tonnes.

Personnel: 8700 total navy; 6800 total air force.

Remarks: The development of the Colombian naval air arm continues slowly because the nation is not able to divert the funds necessary to create a full infrastructure. Recent internal problems, including continued drug smuggling has meant that non essential programmes have been cut back. Nevertheless, the development of the helicopter force, which began in 1983, continues with West German assistance.

MBB BO 105CB helicopters are embarked in FS 1500 type frigates.

Costa Rica

Organisations: Costa Rican Civil Guard.

Organisational structure: As there are officially no armed forces, the President is commander-in-chief of the public security forces which includes an air unit.

Command structure: See above.

Air-capable ships: None.

Shore bases: Limon; Puntarenas.

Embarked aircraft: None.

Shore-based aircraft: Cessna 337 Skymaster (3).

Units: Civil guard air unit.

Recent operations: None.

Typical deployments: None.

Weapon systems: Cessna armed with 7.62 mm machine gun pods underwing and has provision for various rocket systems.

Personnel: 2000 total civil guard.

Remarks: Despite its position in Central America, Costa Rica has found no need of operational forces since the abortive invasion by rebels in 1955. The twin-engined reconnaissance aircraft have other duties besides naval operations. No plans have been announced for further expansion of this element of the civil guard.

Cuba

Organisations: Cuban Revolutionary Air Force.

Organisational structure: Authority passes from the President, as C-in-C through the Minister of the Revolutionary Armed Forces to the individual general staffs.

Command structure: Modelled on Soviet lines with the regiment as the basic operational unit.

Air-capable ships: None.

Shore bases: Camaguey; San Antonio, Santa Clara.

Embarked aircraft: None.

Shore-based aircraft: Mil Mi-4 'Hound' (25); Mil Mi-14 'Haze A' (14).

Units: No details available.

Recent operations: Maritime aircraft have not been involved in any insurgency operations abroad.

Typical deployments: None.

Weapon systems: Soviet type ASW torpedoes, depth bombs and rockets are fitted to all helicopters; typically Mi-14 carries two ASW torpedoes, a dipping sonar and a MAD 'bird'; depth bombs and mines can also be carried.

Personnel: 20 000 total air force.

Remarks: Certainly within the Soviet Bloc, Cuba is a base for USSR long-range maritime reconnaissance aircraft covering the South and mid Atlantic, Caribbean and sea lane approaches to the United States. Maritime reconnaissance is probably undertaken by air force fixed-wing aircraft or by the Soviet MPAs; there are apparently no plans to enlarge the maritime division cover by Cuban forces.

Cyprus

Organisations: National Guard.

Organisational structure: Reports to Greek-Cypriot government.

Command structure: Follows police lines, air unit subordinate to a Colonel (Superintendent).

Air-capable ships: None.

Shore bases: Larnaca, Parvos.

Embarked aircraft: None.

Shore-based aircraft: Agusta-Bell 47G-3 Sioux (2); Pilatus Britten-Norman Maritime Defender (1).

Units: Police Air Unit.

Recent operations: None.

Typical deployments: Coastal patrols to the Green Line demarcation with Turkish-held Northern Cyprus.

Weapon systems: Machine gun and rocket pods can be carried by Defender.

Personnel: Not available.

Remarks: Further developments expected as the division of Cyprus seems likely to continue indefinitely. Cyprus Defence Force received four SA 342 Gazelle helicopters armed with Euromissile HOT in 1988; two more on order; no known naval role.

BN-2 Defender, Cyprus National Guard, seen prior to delivery and showing its secondary role of target-training.

Denmark

Organisations: Royal Danish Navy, Royal Danish Air Force.

Organisational structure: Following the armed forces reorganisation in 1982, all three services are administered by a single defence staff.

Command structure: The helicopter force is limited to eight aircraft by law, aircrew being selected from the Royal Danish Navy and support provided by the Royal Danish Air Force. Squadron is the basic unit, run on UK RAF lines.

Air-capable ships: 5 *Hvidbjornen* class patrol ships.

Shore bases: Vaerlose.

Embarked aircraft: Westland Lynx Mk 80 (7); Westland Lynx Mk 91 (2).

Shore-based aircraft: Gulfstream SMA-3 Gulfstream III (3); Sikorsky S-61A-1 Sea King (7).

Units: RDN: none; RAAF: Esk 721 (Gulfstream III); Esk 722 (Lynx (navy aircrew) S-61A).

Recent operations: None.

Typical deployments: The Gulfstream patrol aircraft operate as far afield as Greenland on deployment, and the embarked helicopters operate off Greenland and in Danish waters, where the S-61As provide SAR facilities for the RDAF tactical aircraft assigned to NATO.

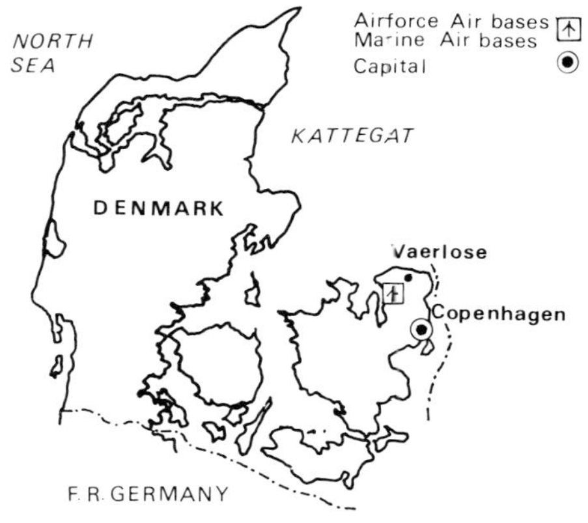

Weapon systems: Helicopters unarmed.

Personnel: RDN: 8300 (total); RAAF: 9000 (total).

Remarks: This small force is regulated by the Danish parliament and although the NATO Council would like a greater effort made by the country to contribute to the Alliance, this seems unlikely. In January 1988, it was announced that the S-61A fleet would be upgraded for night/adverse weather operations rather than be replaced. In 1988, GEC Avionics won orders to fit a FLR system for S-61A fleet.

The Royal Danish Navy operates a small number of Aerospatiale SA 316B Alouette III helicopters for SAR and liaison tasks; a replacement is overdue. *(Robin Walker)*

Sikorsky S-61A, weather-radar equipped helicopter. Denmark will upgrade the performance and capabilities of these aircraft between 1988–90.

The Westland Lynx Mk 80 provides the Danish Coastguard service with a helicopter platform and is operated by the Royal Danish Navy with air force support. *(Westland)*

Dominican Republic

Organisations: Dominican Air Force.

Organisational structure: Republic's president is the commander-in-chief, under him there is political control, although the Under Secretary of State for Aviation is a senior officer. There is no naval aviation function.

Command structure: The squadron is the basic operational unit, modelled on US lines.

Air-capable ships: None.

Shore bases: Santo Domingo.

Embarked aircraft: None.

Shore-based aircraft: Aerospatiale SA 316B Alouette III (2); Cessna T-41D (10).

Units: No details available but thought helicopters operated by one squadron and T-41Ds under control of the military flying school.

Recent operations: None.

Typical deployments: None.

Weapon systems: Aircraft unarmed.

Personnel: 3500 total air force.

Remarks: There is no dedicated maritime patrol function within the Republic's armed forces, other than the use, on a regular but non-specific basis, of helicopters and training aircraft.

Ecuador

Organisations: Ecuadorean Naval Aviation.

Organisational structure: The small naval aviation command was set up to operate shore-based aircraft.

Command structure: Naval aviation: divided into four flights.

Air-capable ships: 1 *Gearing* (FRAM I) class destroyer, 6 *Esmeraldas* class corvettes, 1 *Charles Lawrence* class fast transport.

Shore bases: Guayaquil.

Embarked aircraft: Aerospatiale SA 316B Alouette III (2).

Shore-based aircraft: Beech Super King Air 200T (1); Beech T-34C Turbo-Mentor (3); Bell 206 B JetRanger (5); Cessna Citation I (1); Cessna 320E (1); Cessna 337 Skymaster (3); IAI Arava (1).

Units: naval: 1 Flt (liaison); 2 Flt (T-34C); 3 Flt (Alouette III); 4 Flt (transport).

Recent operations: None reported.

Typical deployment: The Alouette IIIs are used aboard the fast transport ship and occasionally for deck landing practice on the corvettes.

Weapon systems: None.

Personnel: Unknown.

Remarks: The six CNR-built corvettes are the subject of a light helicopter competition, which was apparently won in 1986 by the Westland Lynx but sources indicate that funding has prevented an order for six helicopters.

One of two SA 319B Alouette III search and rescue helicopters which occasionally embark in Ecuadorean Navy corvettes; the helicopters remain in the air force inventory. *(Aerospatiale)*

Egypt

Organisations: Egyptian Navy; Egyptian Air Force.

Organisational structure: navy: basic unit is the Squadron; air force: based on the Regiment, Soviet-style, although the air force has been westernised in recent years.

Command structure: No details available but it is understood that each air force regiment is divided into two or three subordinate squadrons.

Air-capable ships: None.

Shore bases: navy: Port Said.

Embarked aircraft: None.

Shore-based aircraft: navy: Aerospatiale SA 342L Gazelle (12); Mil Mi-8 'Hip C' (14), Westland Sea King Mk 47 (5); air force: Grumman E-2C Hawkeye (5); Ilyushin Il-28 'Beagle' (5); Tupolev Tu-16 'Badger G' (12).

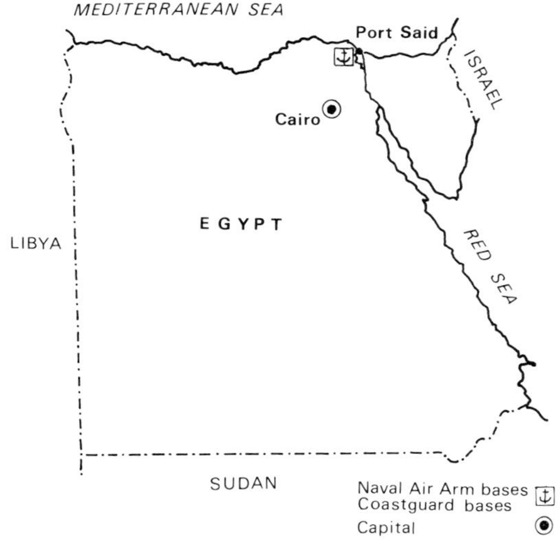

Units: No details available.

Recent operations: No significant operations reported.

Typical deployments: Helicopter and fixed-wing patrol aircraft are flown over the Red and Mediterranean Seas on a regular basis, mainly supporting surface and sub-surface naval operations.

Weapon systems: Navy: all helicopters except Gazelle and Sea Kings are unarmed, latter carry Mk 46 Mod 0 lightweight torpedoes and locally-produced depth bombs. Gazelles are fitted with Euromissile HOT wire-guided ATGW (anti-tank guided weapon) for anti-patrol boat and marine assault duties; air force: 'Beagles' carry full suite of iron bombs and air-launched torpedoes of Soviet pattern. 'Badger' aircraft were originally flown by Soviet crews and current anti-ship missile capability must be limited. Aircraft still have provision for limited ECM (possibly modified Italian) and carry five 23 mm cannon.

Personnel: navy: 30 000 (total); air force 30 000 (total).

Remarks: Although a major westernisation programme has been underway for some time, emphasis in the armed forces has been given to battlefield weapons. Main role of the Hawkeye AEW&C aircraft is to direct fighter aircraft but it is thought that they will be deployed to support naval operations as well. The Sea Kings will have to be replaced by mid-1990s and the Il-28 'Beagles' even sooner, but no funds appear available at present. Tu-16s will not be replaced according to current information. In October 1988, there was reported to be renewed interest in the EH 101 for the Sea King replacement role, probably made under licence.

Ethiopia

Organisations: Ethiopian Air Force.

Organisational structure: Thought to be organised along Soviet lines.

Command structure: No information available.

Air-capable ships: None.

Shore bases: Asmara.

Embarked aircraft: None.

Shore-based aircraft: Agusta-Bell 204B (5); de Havilland Canada DHC-6 Twin Otter (3).

Units: No information available.

Recent operations: Air force maritime units have been involved in anti-Eritrean guerilla operations and in support of visiting Soviet warships.

Typical deployments: Offshore patrol and protection of EEZ with limited offensive capability.

Weapon systems: Aircraft unarmed.

Personnel: 3500 (total).

Remarks: Emphasis remains on battlefield weapon operation and procurement; serviceability of the AB 204B helicopters is doubtful and the exact status of the Twin Otters is not known. There are more than 50 transport aircraft in the air force inventory but all are committed to the anti-separatist operations in Eritrea.

Finland

Organisations: Finnish Air Force; Finnish Coast Guard Frontier Guard.

Organisational structure: air force: organised into wings by geographical deployment and squadrons by type/role with flights deployed around the country; coast guard: operates helicopters in one unit.

Command structure: air force: aircraft squadrons are subordinate to wings, which themselves fall under the authority of the regional defence commands; coast guard: helicopters operated in support of regional units.

Air-capable ships: None.

Shore bases: Kaauhava, Pori, Turko, Uleaborg (Dulu), Utti.

Embarked aircraft: None.

Shore-based aircraft: air force: Aerospatiale AS 332B Super Puma (2) Gates Learjet 35A (3); coast guard: Agusta AB 412 Griffin (1), Agusta AB 206B JetRanger (3), Mil Mi-8 'Hip-C' (2), Piper PA-31 Navajo (2).

Units: air force: Transport Sqn (Learjet); coast guard: no data available.

Recent operations: None reported.

Typical deployments: coast guard: patrol sea approaches to Finland and also supports Frontier Guards with helicopters; recent acquisition of Racal ASR 360 radar for the Mi-8 helicopters has led to an increased capability.

Weapon systems: aircraft: unarmed.

Personnel: air force: 3000 (total); coast guard: 3500.

Remarks: Finland's neutral status and Treaty of Friendship with the Soviet Union limits the size of the armed forces and the type of aircraft which can be acquired. Although many of the fixed-wing combat and transport aircraft could be used for maritime patrol, only the multi-role Learjets have this task at present. Procurement is limited by funding as well as treaty obligations.

Two Finnish Coast Guard Mil Mi-8 'Hip' helicopters, equipped with Racal ASR 360 radar for frontier and coastal patrol. *(IPMS Finland)*

France

Organisations: Naval Aviation (Aeronavale).

Organisational structure: naval aviation: organised into 16 front-line squadrons (flotilles) identified by the suffix F and 18 second-line, training and overseas units, with the suffix S. Front-line units serve with Aviation Embarquee (embarked aviation) or Patrouille Maritime (maritime patrol). Support units are attached to the naval commands at Brest, Cherbourg, Lorient and Toulon. Considerable research is carried out by the Aeronavale.

Two strike carriers, such as FNS *Clemenceau* here, are the pivot of French naval power, carrying nuclear-equipped Super Etendard strike aircraft and F-8E(FN) Crusader fighters. Under current plans, they will be replaced in the next decade by two nuclear-powered carriers.

Command structure: naval aviation: two Flag Officers are responsible for embarked aviation (ALPA) and maritime patrol (ALPATMAR). There

The Super Etendard is capable of conventional iron bomb strike tasks (aircraft on right) or can carry the new ASMP nuclear weapon with a range of 100 km. Note the plane-guard Alouette III helicopter.

The Super Etendard can carry the MATRA Magic short-range air-to-air missile.

is a support organisation which provides training, communications and other specialist equipment to the front-line commands. Each squadron commanded by senior Lieutenant Commander or Commander and general compliment is 12 aircraft. Fixed-wing and helicopter assets are embarked in ships, under ALPA's control, as operational requirements dictate.

Air-capable ships: 2 nuclear-powered aircraft carriers (*Richelieu*, *Charles de Gaulle*) (building), 2 aircraft carriers (*Clemenceau*, *Foch*); 1 helicopter carrier (*Jeanne d'Arc*), 6 *Georges Leygues* class Type C70ASW destroyers (1 building), 2 Type C70AA guided missile destroyers (building), 3 *Tourville* class destroyers, 1 Type 56 destroyer, 1 Mod Type 53 destroyer, 4 FL 3000 class frigates (building), 2 Type TCD 90 assault ships (building), 2 *Duragan* class landing ships, 3 *Argens* class landing ships, 4 *Champlain* class landing ships, 1 guided missile trials ship, 1 underwater research ship, 1 oceanographic research ship, 5 *Rhin* class survey ships, 5 *Durance* class replenishment ships, 1 small nuclear weapons test ship (building).

Shore bases: Aspretto (Corsica), Hyeres, Landivisiau, Lanveoc-Poulmic, Lann-Bihoue, Nimeds-Garons, Noumes (New Caledonia), Papeete (Tahiti), Paris (Le Bourget/Dugny), Rochefort, St Mandrier, St Raphael.

Embarked aircraft: Aerospatiale SA 316B/319B Alouette III (31); Aerospatiale SA 321G Super Frelon (17); Aerospatiale AS 350L Ecureuil (22)*; Aerospatiale SA 365F Dauphin (3); AMD-BA Alizé (21); AMD-BA Etendard IVP (11); AMD-BA Super Etendard (61); LTV F-8E(FN) Crusader (23); Westland Lynx Mk 2/4(FN) (36).

Shore-based aircraft: Aerospatiale SE 3130 Alouette II (13), AMD-BA Alizé (7), AMD-BA Atlantic (29);

Air superiority remains in the hands of the F-8E(FN) Crusader fighter which is scheduled to be replaced by the Rafale in the mid 1990s. *(Paul Beaver)*

Rafale B for naval service, in artist's impression form.

AMD-BA Atlantique 2 (42)*; AMD-BA Etendard IVM (10); AMD-BA Gardian (5); CAP 10B (6), Dassault-Breguet Falcon 10MER (7), Embraer Xingu (16), Fouga Zephyr (12), Morane-Saulnier Paris (9), Nord 262A (10), Nord 262E Fregate (12), Piper PA-31 Navajo (12), SOCATA Rallye 100S (15); SOCATA 235 (14).

Units: embarked aviation: 4F (Alizé), 6F (Alizé); 11F (Super Etendard); 12F (F-8E FN Crusader); 14F (Super Etendard), 16F (Etendard IVP); 17F (Super Etendard), 31F (Lynx) 32F (Super Frelon), 33F (Super Frelon), 34F (Lynx), 35F (Lynx); maritime patrol: 21F, 22F, 23F, 24F (Atlantic); support aircraft: 2S (Nord 262/Navajo), 3S (262/Navajo/Falcon 10MER), 9S (Gardian), 12S (Gardian), 20S (Alouette II/III/Lynx/Super Frelon), 22S (Alouette II/III), 23S (Alouette II/III); training: 50S (Rallye), 51S (CAP 10B), 52S (Xingu), 55S (Nord 262), 56S (262), 57S (Falcon/Paris), 59S (Alizé/Etendard IVM/Zephyr.

Recent operations: In 1984, both *Foch* and *Clemenceau* (prior to recent refit) were deployed to the eastern Mediterranean Sea in support of the Multi-

Now mainly shore-based, the Breguet Alizé anti-submarine warfare aircraft is capable of carrying nuclear depth bombs and air-launched torpedoes, as well as rockets.

See landing on the new landing ship, FNS *Bougainville*, this Lynx Mk 4(FN) of squadron 34F from BAN Lanveoc-Poulmic is operated in the search and rescue role; the bulk of the Lynx fleet is equipped for anti-submarine operations.

National Peacekeeping Force off Lebanon; from 1987 there has been a continued presence in Indian Ocean increased in August/September with deployment of Clemenceau group to Gulf area; in January 1988, there was increased speculation that a Cape-route to the Indian Ocean would be used to support ALINDIEN (Flag Officer Indian Ocean).

Typical deployments: Since 1974, aircraft carrier task groups have been based in the Mediterranean; long-range maritime patrol aircraft are equally divided between Atlantic and Mediterranean deployment; Gardian aircraft are deployed in the Caribbean and New Caledonia, Falcon 10MER to the Pacific (French Navy only user).

Weapon systems: Nuclear role: until 1988, 11F and 17F are capable of carrying the AN 52 tactical nuclear weapon, to be replaced by the Aerospatiale ASMP stand-off weapon for which 55 Super Etendards have been modified; other Super Etendard update features are under review; anti-shipping: Lynx Mk 2/4(FN) helicopters can be armed with the AS 12 wire-guided missile but main tasking is made on Super Etendard with a variety of weapons, including the ASMP, iron bombs, Aerospatiale AS 20, AS 30, AS 37 Martel (also available for Atlantic) and AM 39 Exocet; anti-submarine warfare: France probably has a nuclear depth bomb, similar to the UK WE 177, in her inventory. Conventional ASW weapons include Mk 44 Mod 2 (modified by Thomson to L4 standard) shallow-water and Mk 46 Mod 0 torpedo, the DCN Murene is due to enter service in 1988/89; air defence: Super Etendard/Crusader armed with Matra R 530 and Matra-assembled AIM-9G Sidewinder. ASW sensors include sonobuoys and Thomson-Sintra processing for Atlantic/Atlantique 2, Thomson-Sintra DUAV 4 in Lynx. Etendard IVP has five Omera cameras for reconnaissance.

Personnel: 13 000.

Remarks: Aircraft carrier building programme plans

The Aerospatiale SA 321G Super Frelon has been re-roled from anti-submarine warfare to fleet support and assault. In the new role, the Super Frelon is equipped with the Omera ORB-31 radar and the same Doppler navigation system as the Lynx. *(A Sheridan-Duplaix)*

Maritime reconnaissance tasks in the Pacific Ocean territories is carried out by the AMD-BA Gardian, seen here prior to delivery. Note the large observation window.

Richelieu (1996) and *Charles de Gaulle* (1998) to replace existing CVs, each with a 40-aircraft air group which will include Avion de Combat Marine (ACM), probably the Rafale B aircraft. Until then the Super Etendard will be updated with new Electronique Serge Dassault Anemone radar between 1990-94, supplementing the F-8E(FN) which will be replaced by ACM by 1993. There are no definite plans to replace the Alizé aircraft but discussions about an embarked airborne early warning fixed-wing aircraft seem to have been abandoned in favour of shore-based, air force controlled, Boeing E-3 AWACS

Long-range maritime reconnaissance remains in the hands of the AMD-BA Atlantic, operating from Atlantic Ocean and Mediterranean Sea bases. Because of its age, the aircraft will be retired in the early 1990s.

The Atlantique 2 (formerly ANG) is the replacement for the Atlantic and will enter service in 1990.

aircraft, Long-range maritime patrol updates include the procurement of ten (with a requirement for 24) Dassault-Breguet Atlantique 2 LRMP aircraft with the first unit forming in late 1990 after the first airframe deliveries in mid-1989. *Foch* under refit 1987-89 to embark ASMP-capable Super Etendard strike aircraft. In 1988, two Aerospatiale SA 365F helicopters were procured and the AS 350L1 Ecureuil has been selected to replace the Alouette II but no orders have been announced. The Lynx helicopters are being progressively upgraded to Mk 4 standard with the Rolls-Royce Gem Mk 40 series engine.

Navigation training and VIP communications tasks are carried out by the Falcon 10.

The French Navy bought the Brazilian EMB-121A1 Xingu II for navigation training. *(EMBRAER)*

Future Programmes

Atlantique 2: The improved Atlantic maritime patrol aircraft will enter service 1989-94 in a programme co-ordinated by SECBAT using expertise from Belgium, West Germany, France and the United Kingdom. The aircraft will have a life of 30 years, an improved mean time between overhaul and better weapons carriage. The French Navy has a reported requirement for at least 42 with additional numbers to be ordered by 1990. The Atlantique 2 can carry up to eight Honeywell Mk 46 lightweight torpedoes or two AM 39 Exocet anti-ship missiles. Sensors include the Thomson-CSF Iguane radar, Crouzet MAD and Thomson-CSF Arar 13A radar warning receiver. The total sortie time is 12.5 hours and the maximum endurance could be as great as 18 hours with no reserves.

Crusader replacement: Although considerable publicity has been given to the replacement of the F-8E(FN) by the International Hornet, it is now certain that the Rafale B is the chosen type; this aircraft will also replace the remaining Etendard IVPs in the mid 1990s. The Rafale B will incorporate reinforced main landing gear, a modified nose wheel and the possible use of a mini ski-jump ramp from the new nuclear-powered aircraft carrier design.

NH90: The French Navy will replace its existing Lynx Mk 4(FN) anti-submarine helicopters with the naval version of the multi-nation NH90 project. It is understood that 40 ASW variants are required and that 20 commando assault helicopters of almost similar design will be ordered to replace the remaining Super Frelon helicopters in the late 1990s.

SA 365F Dauphin: In February 1988, it was announced that the French Navy would acquire three SA 365F helicopters immediately to replace the Alouette III plane-guard helicopters used aboard the aircraft carriers, FN Ships *Foch* and *Clemenceau.* It is expected that further orders will be placed for the Dauphin in due course.

Super Etendard/ASMP: The total fleet of Super Etendard naval strike aircraft has been modified to take the nuclear-armed stand-off anti-shipping weapon, ASMP. Xingu II: two more aircraft are on order for communications duties.

Zephyr: To be replaced by the deck-landing capable Alpha-Jet modification according to AMD-BA representatives at Farnborough Air Show 88. Models show a twin-wheel, strengthened nose wheel arrangement.

Gabon

Organisations: Gabon Air Force.

Organisational structure: Follows French lines with the Escadrille de Transport (transport squadron) operating the single maritime patrol aircraft. There is no embarked aviation.

Command structure: Details are not available.

Air-capable ships: None.

Shore bases: Libreville, Port Gentil.

Embarked aircraft: None.

Shore-based aircraft: Embraer EMB-111 Bandeirante (1).

Units: Esc de Transport.

Recent operations: None reported.

Typical deployments: Coastal surveillance and EEZ protection.

Weapon systems: EMB-111 carries 70 mm and 127 mm rocket pods by Aviàbràs and FN Herstal. The primary sensor is the APS-128 search radar; searchlight and limited ECM equipment.

Personnel: 400 (total).

Remarks: A small but efficient air force with modern equipment purchased with the wealth of the country's mineral reserves. With priority being given to the upgrading of the flying training organisation, future maritime patrol and embarked aviation procurement is very unlikely.

EMBRAER EMB-111 Bandeirante maritime patrol aircraft, generally operated unarmed but capable of carrying 127 mm rockets. Standard equipment includes the Litton AN/APS-128 rescue/weather radar.

Germany, East

Organisations: Volksmarine (naval aviation).

Organisational structure: Organised into two air regiments with three naval air stations. Each geschwaller is divided in three squadrons.

Command structure: Follows Soviet lines but with traditional German naval ranks.

Air-capable ships: None.

Shore bases: Laage, Parow Putnitz.

Embarked aircraft: None.

Shore-based aircraft: Mil Mi-8 'Hic' (16); Mil Mi-14 K 'Haze A' (8); Su-22 'Fitter' (20).

Units: *Kurt Barthel* (helicopters); MFG-28 (Su-22).

Recent operations: None recorded.

Typical deployments: Operations confined to the Baltic Sea in support of Soviet, Polish and own naval operations. Naval infantry assaults are carried out by the Mi-8s, which are also used for SAR.

Weapon systems: Soviet-designed but locally built ASW torpedoes, depth bombs and other ASW equipment, including lightweight dipping sonar and sonobuoys for the Mi-14. Su-22 is equipped for anti-ship operations in the Baltic Sea. Mi-8 helicopters can be armed with AT-2 'Swatter' guided missiles, and 55 mm or 68 mm rocket pods.

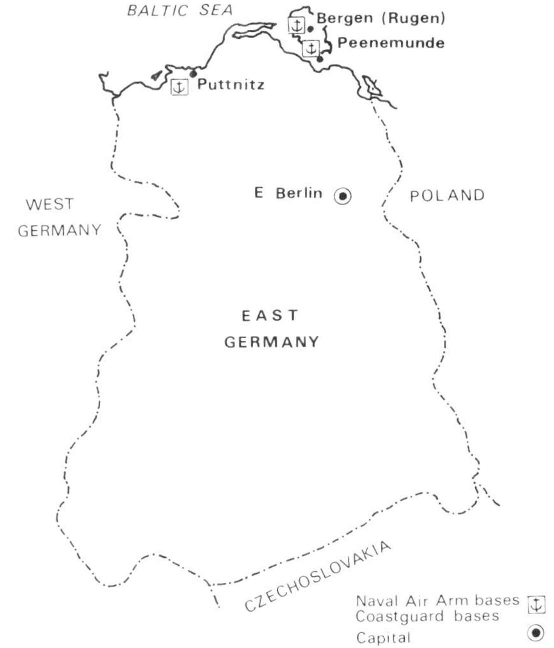

Personnel: No data available.

Remarks: None.

East German naval aviation has recently been equipped with the Mi-14 'Haze A' land-based anti-submarine helicopter.

Germany, West

Organisations: Marineflieger (naval aviation).

Organisational structure: The basic operational unit is the squadron, three of which form a wing (Marinefliegergeschwader); two wings are equipped for strike/maritime reconnaissance and two for helicopter operations. Aircrew training is undertaken by the Luftwaffe (air force) in the United States.

Command structure: The Marineflieger has over 130 combat aircraft and four naval air stations, commanded by a flag officer. Embarked Sea Lynx flights are complemented for 3 pilots/2 CPO observers (one helicopter) or 5 pilots and 3 observers (two); maintenance support is an air engineering officer and nine ratings.

Air-capable ships: 8 *Bremen* class frigates.

Shore bases: Eggebek; Kiel-Holtenau; Nordholz; Schleswig-Jagel.

Embarked aircraft: Westland Sea Lynx Mk 88 (19).

Shore-based aircraft: AMD-BA Atlantic 1 (14); AMD-BA Atlantic ELINT (5); Dornier Do28D-2 (19); Dornier 228 (1); Panavia Tornado IDS (92); Westland Sea King Mk 41 (21).

Units: fighter: MFG-1 (Tornado); MFG-2: (Tornado); helicopter ASW: MFG-3 (Sea Lynx); helicopter SAR: MFG-5 (Sea King/Do28D); maritime patrol MFG-4 (Atlantic).

Recent operations: Under Federal German law, military forces have limited deployments outside the NATO commitment. Helicopter-equipped frigates operate with STANAVFORLANT (Standing Naval Force Atlantic) and have been supporting the increased European NATO presence which has supplemented the detachment of US Navy warships to the Gulf.

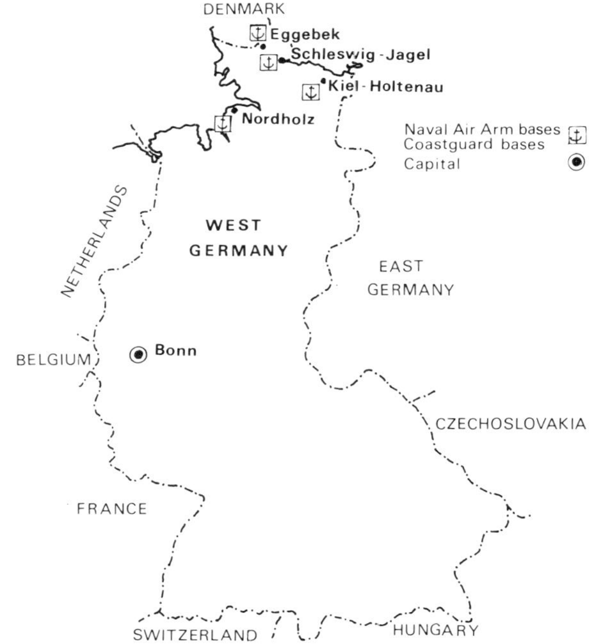

A Panavia Tornado of the West German Navy's MFG-1, armed with four Texas Instruments HARM anti-radiation missiles and ECM pods. *(MBB)*

Typical deployments: Embarked aircraft operate from host warships for NATO exercises and other training activities, including a close working relationship with the UK Royal Navy. Shore-based aircraft have been deployed around the European seaboard and to Iceland and other Atlantic islands. Generally, the maritime air forces keep within the confines of the North and Baltic Seas. In the 1990s, the West German Navy expects to have 24 helicopter-capable frigates in commission and has therefore reduced its minimum operational requirement to one helicopter.

Weapon systems: For anti-ship operations, the Tornados can be armed with the MBB Kormoran missile. Sea King Mk 41 helicopters are receiving the British Aerospace Sea Skua missile, integrated with the Ferranti Sea Spray Mk 3 radar. The Sea Lynx is armed with the Honeywell Mk 46 Mod 1 and the locally modified Mk 44 for shallow water ASW operations; the standard sensor is the Bendix AQS-

Equipped with a MBB reconnaissance pod, this is a Tornado of MFG-2, part of the West German Navy's commitment to NATO. Picture taken in 1987.

18 lightweight dipping sonar. Five of the Atlantics are configured for ELINT (Electronic Intelligence) operations. ASV Atlantics carry two AS 30 missiles.

Personnel: 6000.

Remarks: The Tornado programme was complete by late 1988 as the F-104G Starfighter had been phased out of service in 1986/7. Future plans include the replacement of the Atlantic aircraft (MPA 90) in 1992–96 with the Lockheed P-3G LRAACA, modified by MBB with the Dornier; the AMD-BA Atlantique ATL2 is the outside contender. Helicopter procurement includes plans to replace the existing Sea King and Do 28D force with 72 medium high technology helicopters, probably the naval version of the NH 90. The single Dornier 228 has been under evaluation. The Tornado force will receive the Kormoran 2 from 1989 and already carries twin Mauser 27 mm cannon and up to four AIM-9L Sidewinder air-to-air missiles.

The standard small ship's helicopter is the Westland Sea Lynx Mk 88 helicopter, equipped with dipping sonar and ASW torpedoes. *(Paul Beaver)*

Considerable efforts have been made to integrate the Sea Skua missile and Ferranti Sea Spray Mk III (not fitted here) as part of the West German Navy's programme to increase the Sea King's anti-ship capability.

Future programmes

Atlantic Combat Value Enhancement: After 15 years of service, the Breguet 1150 Atlantic maritime patrol aircraft force of the West Germany Navy's MFG-3 at Nordholz has undergone a major integration programme. Known as the Combat Value Enhancement programme, the programme is under contract to Dornier. The following sub-systems have been improved: radar (AN/APS-134(V)1), acoustics (AN/AQA-5 Mod), navigation (LN 33 with Decca), electronic surveillance measures (EW 1017A in wing-tip pods), sonobuoy launcher (Dornier), acoustic recorder system (M14L). Test installations were undertaken in 1980 and work began on a progress basis in 1984. The last airframe update was completed in September 1987.

Lynx replacement: The West Germany Navy is looking for a new helicopter to equip the proposed NFR90 frigate programme and until October it was thought that the multi-national NH 90 would be selected. Concern now centres on the cost of the helicopter and the IOC could be delayed until 2000. The first helicopters could be delivered in late 1996 and there is a reported requirement for 72 land-based and shipborne helicopters; in addition, the West German Luftwaffe has a reported requirement for 60 NH 90s for the combat SAR and specialist transport role.

Sea King improvement: To enhance the operational capability of the Sea King helicopter force, MBB has been awarded an unspecified value contract to integrate anti-ship missile and combat rescue systems. The West German Navy selected the Ferranti Sea Spray Mk III radar for a dual installation to give 360 degrees cover and to allow integration with the British Aerospace Sea Skua anti-shipping missile. In addition, various unspecified electronic warfare systems have been fitted to the helicopter. Work is reported in *Jane's Defence Weekly* to be progressing slowly.

Tornado Update: The strike capability of the aircraft was improved in 1988 with the entry into service of the Kormoran 2 anti-shipping missile. In addition, a new EW fit has been procured and it is reported that engine modifications are underway. The naval Tornado will also carry the Texas Instruments HARM anti-radiation missile. A total of 110 Tornado aircraft will be acquired for front-line units and war reserves.

Ghana

Organisations: Ghana Air Force.

Organisational structure: Organised into two combat squadrons, one of which operates the single maritime patrol aircraft. There is no embarked aviation.

Command structure: A single squadron operates the two F27s.

Air-capable ships: None.

Shore bases: Accra.

Embarked aircraft: None.

Shore-based aircraft: Fokker F27 Mk 400M (2).

Units: No 2 Sqn (F27).

Recent operations: None reported.

Typical deployments: The air force is predominantly an internal security force with limited combat equipment.

Weapon systems: Simple depth bombs may be carried but the aircraft are thought to be unarmed.

Personnel: 1400 (total).

Remarks: A limited force because of the country's economic situation; it is unlikely to grow in the foreseeable future.

Greece

Organisations: Hellenic Naval Aviation; Hellenic Air Force.

Organisational structure: The navy has operational control land-based aircraft belonging to the air force, operating helicopters only in naval colours. Two squadrons are in commission.

Command structure: naval aviation: the two squadrons report to a flag officer naval aviation and are known as the air wing; air force: the squadron is the basic unit with two or more making up a wing.

Air-capable ships: 1 *Allen M Sumner* (FRAM II) class destroyer; 1 *Gearing* (FRAM II) class destroyer; 2 *Kortenaer* class frigates; 1 *Naftilos* class survey ship; 1 training ship; 1 landing ship.

Shore bases: Eleusis, Mitilini, Pireffs, Thessaloniki (Mikra).

Embarked aircraft: naval aviation: Aerospatiale SA 319B Alouette III (4), Agusta AB 212ASW (18).

Shore-based aircraft: air force: Grumman HU-16B Albatross (8).

Units: naval aviation: No 1 Sqn Alouette III/No 2/3 Sqns AB 212ASW; air force: 353 Sqn (HU-16B).

Recent operations: None reported.

Typical deployments: NATO operations in the Eastern Mediterranean Sea have been overshadowed in recent years by an increased readiness against Turkey.

Weapon systems: Alouette IIIs have been configured for Aerospatiale AS 12 wire-guided missiles, Honeywell Mk 44 Mod 2 torpedoes and depth bombs; AB 212ASWs have bendix AQS-18 sonar, Mk 44/ Mk 46 torpedo and depth bomb capability; HV-16B

carry ASW torpedoes, depth bombs and will be modernised by 1990 if funds permit.

Personnel: Not available.

Remarks: In 1987, Greece issued a request for proposals for 12 long-range ASW helicopters to which the following manufacturers are thought to have responded: Aerospatiale (AS 332F Super Puma), Agusta (ASH-3H Sea King), EHI (EH 101), Kaman (SH-2G Super Sprite), Sikorsky (S-70B Seahawk) and Westland (Advanced Sea King). Greece's small ship ASW helicopter requirement has been met by the AS 212ASW, which is now being considered for an anti-shipping role. Greece is also attempting to find funds for the HU-16B Albatross update first made public at the 1984 Defendory show.

Greece now operates the Agusta-Bell 212ASW as the standard shipborne helicopter for surface search and anti-submarine warfare operations.

Future programmes

Albatross update: Since 1984, there has been considerable interest in the Hellenic Ministry of Defence's proposal to update the Albatross amphibian for service into the next century. The aircraft requires a new radar system (for which the MEL Super Searcher was thought to be favoured), new radar warning receivers and new engines (probably the Garrett TPE331-15 turboprop with Dowty Rotol four-blade propeller.) Sensors could include the GEC Avionics ASQ-902(v) accoustic processor and Italian-designed ECM equipment. No details of the programme's progress are available.

Land-based ASW helicopter: The primary programme currently under discussion is the land-based anti-submarine helicopter for which most naval helicopter manufacturers are competing. In late 1987, the Hellenic Navy indicated to *Jane's Defence Weekly* that the Westland Sea King with the Thomson Sintra HS-12 dipping sonar would be selected but by October 1988 no decision had been made. The EH Merlin helicopter is reported to be the long-term goal of the Hellenic Navy.

Small shipborne ASW helicopter: The Agusta 212ASW helicopter is due to be supplemented in the near future but the helicopter selection rather depends on the ships acquired. If, as is suspected in the United States, some former US Navy *Knox* class frigates are made available, it is possible that the Kaman SH-2G Super Sprite will be acquired. Otherwise a further buy of AB 212ASWs is forecast, but with anti-ship missile capability.

Guyana

Organisations: Guyana Air Command.

Organisational structures: A small air force with no combat aircraft.

Command structure: One integrated unit.

Air-capable ships: None.

Shore bases: Georgetown.

Embarked aircraft: None.

Shore-based aircraft: Beech Super King Air 200 (1), Pilatus Britten-Norman BN-2A Maritime Defender (6), Shorts Skyvan (1).

Units: None.

Recent operations: None reported.

Typical deployments: Mainly operated for coastal patrol linked to internal security operations.

Weapon systems: Aircraft unarmed; all carry light weight search radar.

Personnel: 7000 (total).

Remarks: A limited force which is mainly engaged in liaison and transport tasks.

A pre-delivery photograph of a Guyana government Shorts Skyvan 3M used for coastal surveillance and patrol work. *(Shorts)*

Honduras

Organisations: Honduran Air Force.

Organisational structure: Two combat units fly jet equipment and the Transport Squadron operates the propellor-driven aircraft.

Command structure: No reliable data available.

Air-capable ships: None.

Shore bases: La Ceiba, San Pedro Sula, Tegucigalpa.

Embarked aircraft: None.

Shore-based aircraft: Embraer EMB-111 Bandeirante (2).

Units: Esc de Transporte (EMB-111).

Recent operations: None reported.

Typical deployments: The internal security situation and the risk of invasion from neighbouring countries keeps most of the maritime patrol capability devoted to anti-insurgency.

Weapon systems: EMB-111 carries six to eight Aviàbràs 127 mm rockets or up to 28 rockets (70 mm FFAR) in underwing pods.

Personnel: 1200 (total).

Remarks: Until the border situation with Nicaragua is resolved the bulk of the defence spending will be aimed at internal operations.

Hong Kong

Organisations: Royal Hong Kong Auxiliary Air Force.

Organisational structure: RHKAAF reports to the Governor of the Colony; UK Royal Air Force and UK Army Air Corps also deployed reporting to Governor via the military command structure.

Command structure: The RHKAAF is organised as if it were a department of state in the UK, but everyday operations follow the UK RAF lines.

Air-capable ships: None.

Shore bases: Kai Tak; Sek Kong.

Embarked aircraft: None.

Shore-based aircraft: Aerospatiale SA 365C Dauphin (2); Cessna 404 Titan (1); Pilatus Britten-Norman Islander (1).

Units: RHKAAF has no subordinate structure.

Recent operations: Continued EEZ, anti-smuggler and 'anti-boat people' reconnaissance in support of the Royal Hong Kong Police.

Typical deployments: Continued presence around the 600 plus islands, SAR and internal security operations.

Weapon systems: Aircraft unarmed.

Personnel: 250.

Remarks: British forces provide bulk of the defence of the Colony with Royal Air Force Wessex HC 2 helicopters (28 Squadron) and British Army Scout AH 1 helicopters (660 Squadron AAC) deployed to Sek Kong; gradual turn-over to Hong Kong operations continues in the run-up to Chinese take-over in 1997.

Hong Kong still operates two SA 365C Dauphin helicopters for transportation, police liaison and SAR tasks. *(Paul Beaver)*

Iceland

Organisations: Icelandic Coast Guard.

Organisational structure: The country has no standing armed forces as its defence is guaranteed by the USA; the coast guard carries out maritime patrol and SAR duties.

Command structure: Civilian chain of command.

Air-capable ships: 2 *Aegir* class fishery protection craft; 1 Odinn class patrol vessel.

Shore bases: Reykjavik.

Embarked aircraft: Dauphin capable but not normally embarked.

Shore-based aircraft: Aerospatiale SA 365N Dauphin 2 (1); Fokker F27 Friendship (1).

Units: None.

Recent operations: None reported.

Typical deployments: Police the fishery and natural resource zones around the coast.

Weapon systems: All aircraft unarmed.

Personnel: 200 (total).

Remarks: NATO long-range maritime patrol aircraft are based at Keflavik on rotation from USA, Canada, Netherlands, UK and West Germany.

The sole Aerospatiale SA 365N Dauphin 2 of the Iceland Coast Guard. *(Aerospatiale)*

Longer range surveillance and rescue coordination is carried out by a single Fokker F27.

India

Organisations: Indian Navy; Indian Coast Guard.

Organisational structure: naval aviation: this consists of 12 squadrons; the Chief of the Naval Staff exercises command through three area Flag Officers Commanding-in-Chief, including the Flag Officer Commanding-in-Chief Southern Naval Command at Cochin.

Command structure: The parent command for naval aviation is Southern Naval Command, which includes the subordinate command of Flag Officer Naval Aviation. Service called the Naval Air Force and organised into squadrons and flight along British lines, although units with Soviet equipment are thought to come under some control elements of Soviet advisers. Indian Coast Guard air operations commenced in 1982 and are directed from the headquarters in New Delhi under the auspices of three regional subordinate commanders in Bombay (Western), Madras (Eastern) and Port Blair (Andaman & Nicobar Islands). There is a coast guard district for each of India's maritime states. Training is carried out at the Naval Academy at Cochin, fixed wing aircrew with Indian Airlines and rotary with the navy at Garuda. Dornier 228 training undertaken at Munich.

Air-capable ships: 2 aircraft carriers (*Viraat* & *Vikrant*); 5 'Kashin' class destroyers (1 building); 3 *Godavari* class frigates (3 building); 6 *Leander* class frigates; 1 survey ship; 1 *Urga* class submarine tender; coast guard: 6 *Vikram* class OPVs.

Shore bases: Cochin, Dabolim (Goa); Coast Guard: Calcutta: Dabolim (Goa); Santa Cruz (Bombay).

Embarked aircraft: naval aviation: Aerospatiale Alouette III (11); British Aerospace Sea Harrier FRS 51 (23) (10)*; Kamov Ka-25 'Hormone' (5); Kamov Ka-27 'Helix' (18); Westland Sea King Mk 42A (9); Westland Sea King Mk 42B (20)*; Westland Sea King Mk 42C (6); coast guard: HAL Chetak (6).

Shore-based aircraft: naval aviation: Aerospatiale Alouette III (11); British Aerospace Sea Harrier T 60 (3); Breguet Alizé (5); HAL HJT-16 Kiran (7); HAL Chetak (3); Hawker Sea Hawk FB 5 (2); Hughes 269B

India's new aircraft carrier, *Viraat*, will operate the Sea Harrier FRS51 for combat air patrol and strike duties.

Westland is supplying a number of Sea King helicopters, such as the Sea Eagle missile-armed Mk42B. *(Westland)*

(4); Ilyshin Il-38 'May' (3); Pilatus Britten-Norman Defender (18); SEPECAT/HAL Jaguar International (12); Tupolev Tu-142M 'Bear F' (5) (3)*; coast guard: Dornier 228 (36)*; Fokker F27 Friendship (2); HAL Chetak (5) (3)*.

Units: naval aviation: 300 (Sea Harrier); 310 (Alize); 315 (Il-38); 318 (Defender); 321 (Alouette III/Chetak); 330 (Sea King); 331 (Chetak); 333 (Ka-25); 336 (Sea King); 550 (Defender); 551 (Kiran/Sea Hawk); 552, 561, 562 (Alouette III/Chetak/Hughes 269B); coast guard: 700 (Fokker F27/Dornier 228); 800 (Chetak).

Recent operations: In the 1970s, the Indian Navy was in action against Pakistan naval forces and the

India's Coast Guard is taking delivery of Dornier 228 maritime patrol aircraft, the bulk of which will be built under licence by HAL. *(Pin Point)*

Fokker F27 aircraft continued to be operated in coastal surveillance and patrol duties by the Indian Coast Guard.

main perceived threat is still Pakistan. In 1987, Indian naval forces, including maritime reconnaissance assets were used to patrol the waters around Sri Lanka.

Typical deployments: The fixed-wing aircraft are embarked in the aircraft carrier *Vikrant* which was joined in autumn 1987 by *Viraat* (ex *Hermes*); both carriers will embark the Sea Harrier and Sea King helicopters only, following the last embarkation of the Alizé ASW aircraft in May 1987. Destroyers embark the Ka-25 or the Ka-27; the latter is replacing the former along Soviet lines. The *Leander* class frigates formerly embarked the Alouette III but now operate Sea Kings in the ASW and soon ASV role (with British Aerospace Sea Eagle). The purchase of three more 'Kashin' Class destroyers will mean that more Kamov-built helicopters will be required. The Coast Guard Dornier 228, equipped with MEL Marec 2 radar, will be used for coastal surveillance and limited SAR tasks, supplementing the eleven Alouette IIIs. The larger fixed-wing types, Super Constellation and Il-38 are used for maritime reconnaissance tasks.

Weapons systems: Soviet equipment is armed with Russian designed equipment, but the Westland Sea King Mk 42Bs will have provision for Sea Eagle anti-ship missile; reports that India is acquiring the air-launched Aerospatiale AM 39 Exocet cannot be confirmed but it is understood that an order with BAe for the Sea Eagle missile has been signed and that the weapon will be used on the Jaguar strike aircraft, possibly transferred to the navy from the air force. Sea Harriers are armed with a suite of 30 mm Aden cannon, various 'iron' bombs and will eventually have the Sea Eagle anti-ship missile as well. The Indian Air Force maritime strike Jaguars will also receive Sea Eagle. The lightweight, air-dropped torpedoes for the Alizé and Alouette III ASW aircraft are thought to be of French manufacture.

Personnel: naval aviation: 2000; coast guard: no data available.

Remarks: The last deck launch of the Alizé was performed on 12 May 1987 aboard *Vikrant* but the aircraft will remain in shore-based service until 1990. India is continuing its build up of naval forces, largely with Soviet assistance and has apparent plans to dominate the Indian Ocean to contain US and other 'outsider' powers' influence. India has revived plans to acquire an airborne early warning helicopter to operate from *Vikrant* and *Viraat*. It could acquire up to six Sea King Mk 42 helicopters or request funds to modify existing Sea King airframes, using the Thorn EMI Electronics Searchwater radar with HAL being the main contractor; the AEW version would be designated Sea King Mk 42D. A further six twin-engined helicopters are planned for SAR duties and the Coast Guard plans a total of 30 light helicopters for shipborne operations.

Indonesia

Organisations: Indonesian Air Force; Indonesian Naval Aviation.

Organisational structure: air force: divided into five functional commands, four of which have aircraft allocated; there are five operational squadrons with combat aircraft, one squadron of which is devoted to maritime patrol aircraft; naval aviation: maritime patrol and transport aircraft are organised into four squadrons.

Command structure: air force: organised into squadrons; naval aviation: embarked aviation is organised into flights, detached from parent squadrons and shore-based into squadrons.

Air-capable ships: 4 *Van Speijk* class frigates; 1 *Nala* class frigate; 1 *Hajar Dewantara* class frigate; 3 'Tribal' class frigates; 6 *Teluk Semangka/Tacomca* class landing craft; 1 submarine tender.

Shore bases: air force: Madiun, Semarang; naval aviation: Jakarta (Kemayoran); Surabaya.

Embarked aircraft: air force: naval aviation: Aerospatiale SA 318C Alouette II (3); Aerospatiale

SA 316B Alouette III (3); Bell 47J (4); Nurtanio Aerospatiale NAS-332F Super Puma (6) (20)*; Nurtanio/MBB NBO 105C (20); Westland Wasp HAS 1 (14).

Shore-based aircraft: air force: Boeing 737-200 Surveiller (3); GAF Searchmaster B (12); GAF Searchmaster L (6); Grumman HU-16B Albatross (4); Lockheed C-130H-MP Hercules (1); (2)*; Nurtanio/CASA CN 235 (6)*; Northrop F-5E Tiger II (16); naval aviation: Bell 47G-3 (4); CASA/Nurtanio NC 212 Aviocar (4); CASA/Nurtanio CN 235 (16)*;

The first NAS 332 Super Puma for shipborne duties were delivered to the Indonesian Navy in 1987 and the first helicopter is seen lashed to the deck of *Teluk Banten*, one of six LSTs in service. *(IPTN)*

Douglas C-47 (5); Nurtanio/Bell NB 412 (16)*; GAF Searchmaster B/L (6/12); Piper Cherokee (6).

Units: air force: No 5 (HU-16/C-130H/Boeing 737), No 14 (F-5E); naval aviation; 200 Sqn (NBO 105); 400 Sqn (Wasp); 600 Sqn (C-47/NC 212); 800 Sqn (Searchmaster).

Recent operations: None reported.

Typical deployments: The Indonesian archipelago is as wide as CONUS (Continental United States) and much of the naval and air force maritime patrol effort is put into internal security and patrolling the EEZ. Deployments overseas are rare.

Weapon systems: The Tigers are armed with a variety of ground-attack and anti-shipping cannon and rockets but no missiles; some of the Super Pumas will be configured to carry AM 39 Exocet missiles as well as lightweight torpedoes; embarked helicopters are armed with lightweight torpedoes for ASW; long-range maritime patrol aircraft have a variety of ASW and anti-shipping missiles, but exact details are difficult to confirm.

Personnel: air force: 25 000 (total); 40 000 (total).

Remarks: In August 1987, it became clear that a maritime version of the CN 235 transport was on order for the air force, but other modernisation plans have been hit by the severe economic situation.

Iran

Organisations: Iranian Islamic Air Force, Iranian Islamic Navy, Iranian Revolutionary Guards Corps (equipment frequently changes between three services).

Organisational structure: In Revolutionary Iran it is not possible to truly identify the organisation of the three services which have been described as 'mere shadows' of their former selves.

Command structure: No accurate information is available.

Air-capable ships: 1 *Allen M Sumner* class destroyers, 4 Yarrow-built landing ships, 1 large AOR (*Kharg*).

Shore bases: Bandar Abbas, Bushehr, Kharg Island and various offshore platforms.

Embarked aircraft: Agusta AB 204ASW (12).

Shore-based aircraft: Agusta-Bell 206A JetRanger (6); Agusta-Bell 205A (3); Agusta-Sikorsky ASH-3D Sea King (10); Fokker F27 Troopship (3); Lockheed P-3F Orion (1); Rockwell Shrike Commander (2); Sikorsky RH-53D (9); Lockheed C-130 (5).

Units: No details are available.

Recent operations: In August 1987, the remaining Sea Stallion AMCM helicopters were seen on Western television undertaking minesweeping tasks. In

Only one Lockheed P-3F Orion is thought to remain in service with the Iranian Air Force; this photograph was taken in 1981 when two F-14A Tomcats from USS *America* intercepted the aircraft in the Arabian Sea.

addition, the Iranian Revolutionary Guard Corps has reportedly been operating civilian-registered Bell 212 light support helicopters from oil platforms and other offshore installations to 'spot' possible targets for the 'Boghammer' speedboats. Speculation about the number and types of Iranian helicopters, flown by the 'official' military and IRGC, increases and it is thought that Agusta AB 212ASWs, Bell 212 and Agusta-supplied Sea Kings were operational during the latter part of 1987 and early 1988.

Typical deployments: Maritime aircraft are only operated in the Gulf and in the Arabian Sea approaches to the Strait of Hormuz.

Weapon systems: The ASW helicopters carry the Whitehead A/244S lightweight torpedo and the Orion probably still has the Mk 46 available in limited numbers, but the threat is now surface and air. The transport and training aircraft are unarmed. It is possible that the AB 204AS helicopters are armed with the Aerospatiale AS 12 anti-shipping missile but reports of AB 212 ASW/MARTE cannot be confirmed.

Personnel: No firm data available.

Remarks: The very nature of the society and the recent war situation in the Gulf means that accurate data is extremely difficult to assemble.

Iraq

Organisations: Iraqi Naval Aviation; Iraqi Air Force.

Organisational structure: Details of the order of battle of both air arms are not available but it is understood that there are two armed forces commands which controlled air operations in the war against Iran: Air Defence Command and Support Command. The precise status of the Iraqi Naval Aviation arm is not known.

Command structure: Details of squadrons and other units are not available, but probably run along Soviet lines as the USSR is the major arms and training supplier. French influence is also strong.

Air-capable ships: 4 *Lupo* class frigates (on order - see remarks).

Shore bases: Basrah (Maqal); Shaibah.

Embarked aircraft: None.

Shore-based aircraft: Aerospatiale SA 321H Super Frelon (7); Agusta AB 212ASW (8); Dassault-Breguet Mirage F1EQ5 (72).

Units: No details are available.

Recent operations: Frequently operating strike and reconnaissance sorties over the Gulf to search for Iranian and other merchant ships; in May 1987 one (possibly two) Mirage F1EQ strike aircraft engaged in error USS *Stark*, a US patrol frigate, with two AM39 Exocet missiles and disabled the ship. Mirage strike aircraft were used weekly in the 'tanker war' against Iranian and neutral shipping.

Typical deployments: All operations were concerned with the Iran-Iraq war but now at lower level.

Weapon systems: For anti-shipping strikes, the Aerospatiale AM39 Exocet is carried by the Mirage F1EQ and Super Frelon; AB 212ASWs probably only carry ASW weapons such as the Whitehead A/244S lightweight torpedo.

Personnel: navy: 5000 (total); air force: 35 000 (total).

Remarks: Until 1984, a unit of loaned Super Etendard was in service prior to the delivery of the Mirage F1 aircraft. Super Frelon helicopters modified for Exocet missiles in France in early 1980s. In July 1988, prior to the cessation of hostilities with Iran, an order was signed with Aerospatiale for eight SA 365F Dauphin helicopters fitted with Thomson-CSF Agicon radar and armed with AS 15TT missiles; delivery is expected in 1989/90. Four Lupo class frigates were ordered from Italy but are reported to be retained in Italy for the duration.

Ireland

Organisations: Irish Army Air Corps.

Organisational structure: There is a single helicopter squadron and a maritime squadron allocated to No 1 Support Wing, with three other operational squadrons flying non-maritime tasks.

Command structure: Modelled on British lines to some degree, the IAAC has a command structure reporting to the Minister of National Defence and the Cabinet.

Air-capable ships: 1 P31 class fishery protection craft (*Eithne*).

Shore bases: Baldonnel (Casement), Gormanston.

The most modern equipment in the Irish Air Corps, the controller of all military flying, is the Aerospatiale SA 365F Dauphin helicopter. Five have been procured for embarked and land-based duties. *(Aerospatiale)*

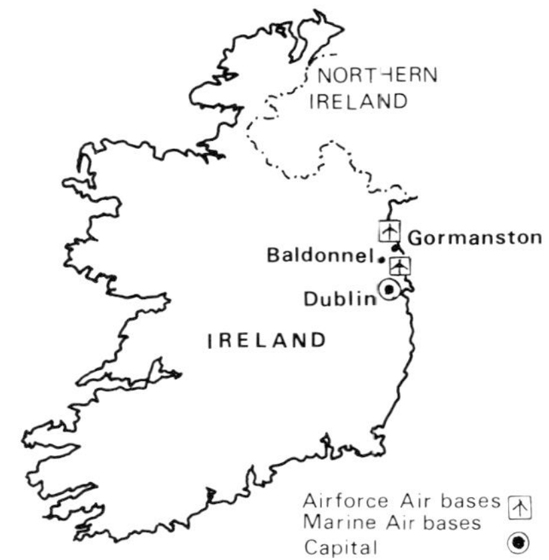

Embarked aircraft: Aerospatiale SA 365F Dauphin 2 (2).

Shore-based aircraft: Aerospatiale SA 365F Dauphin 2 (3); Beech Super King Air 200T (2).

Units: Helicopter Sqn (Dauphin); Maritime Sqn (King Air).

Recent operations: None reported.

Typical deployments: Coastal patrol, fishery protection and SAR.

Weapon systems: Irish military aircraft are not armed but are equipped with Bendix RDR 1500 radar.

Personnel: 800 (total).

Remarks: With the arrival of the French-built helicopters, the only real plans now are for a longer-range maritime patrol aircraft, but there are no funds available.

Israel

Organisations: Israel Defence Force (Air Force).

Organisational structure: Organised to fight a land war and for internal security rather than maritime operations and there are thought to be a total of 18 combat squadrons plus support.

Command structure: No details available.

Air-capable ships: 2 Sa'ar 4 fast missile craft; 2 Sa'ar 4.5 class fast missile craft, 4 Sa'ar 5 corvettes (planned).

Shore bases: Eilat, Haifa (Ramat David), Tel Aviv (Ben Gurion).

Embarked aircraft: Aerospatiale SA 366G Dolphin (2) (20)*.

Shore-based aircraft: Bell 212 (25); Grumman E-2C Hawkeye (4); IAI 1124 Sea Scan (7).

Units: No details available.

Recent operations: Almost monthly anti-terrorist operations along the coast of Israel and Lebanon.

Typical deployments: No details available.

Weapon systems: SA 366Gs could be armed with a medium range anti-shipping missile or with ASW systems.

Personnel: 28 000 (total).

Remarks: There is a requirement for up to 20 shipborne helicopters of the Dauphin type which would probably be procured through US FMS because of the French embargo; the two SA 366Gs currently under evaluation are early prototypes struck off USCG inventory. The Dauphins would be used for fire control, SAR and ASW duties. The future of the Sa'ar 5 corvettes is in doubt and first Sa'ar 4 corvettes may not be delivered until 1998.

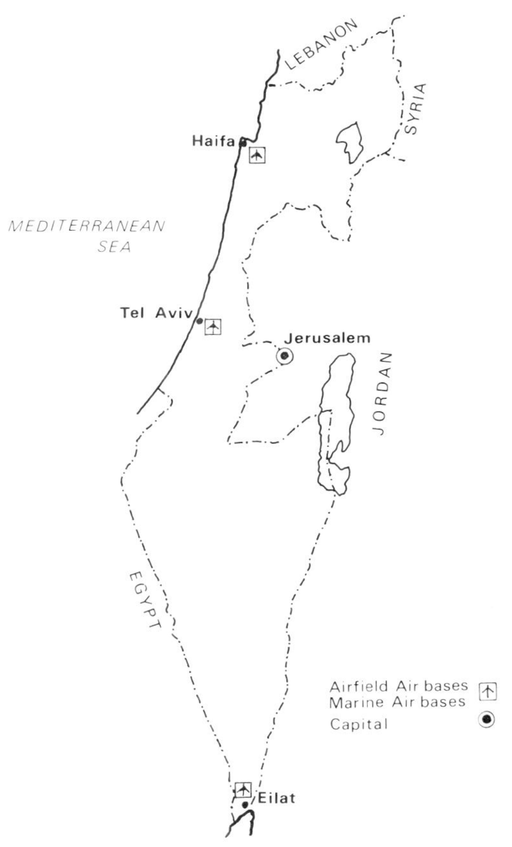

The Israel Air Force provides airborne early warning facilities for naval forces with the Grumman E-2C Hawkeye aircraft. *(IDF)*

Israel Aircraft Industries have delivered an unknown quantity of the Sea Scan version of the Westwind. *(IAI)*

Italy

Organisations: Marinavia (Italian navy aviation).

Organisational structure: The naval air arm operates helicopters but has aspirations to equip with the Sea Harrier or another STOVL project; the long-range maritime patrol aircraft are under naval operational control but owned by the air force. There are now nearly 100 naval helicopters in service.

Command structure: air force: the basic operational unit is the Stormo (regiment), although SAR helicopters are operated in support of specific bases. Naval: the operational command rests with the Director Aircraft (Naval) – Ispettore dell'Aviazione per la Marina (MPA) – currently Air Vice Marshal Vincenzo Manca; the head of the Italian Ministry of Defence Dept 6 (Helicopters) – Capo del 6° Reparto Elicotteri dello Stato Maggiore Marina – is currently (1988) Rear Admiral Umberto Rattigelli; the director general of the coast guard branch – Ispettore Generale dell Carpo delle Capitanerie di Parto – is Rear Admiral (Senior) Francesco Cerenza.

The current air element of the new aircraft carrier *Guiseppe Garibaldi* is the Agusta ASH-3D Sea King; a decision on the procurement of fixed-wing aircraft is awaited.

Air-capable ships: 1 light aircraft carrier (*Garibaldi*); 1 cruiser (*Vittorio Veneto*), 2 *Andrea Doria* class cruisers; 2 *Animoso* class destroyers (building); 2 *Audace* class destroyers; 8 *Maestrale* class frigates; 4 *Lupo* class frigates; 4 *Minerva* class frigates (8 building); 2 amphibious warfare ships (building); 2 *De Soto County* class landing ships; 2 *Stromboli* class replenishment ships; 1 salvage ship.

Shore-bases: naval: Catania (Fontanarossa); La Spezia (Luni); Taranto (Grottaglie); air force: Brindisi, Cagliari (Elmas); Catania; Rimini (Miramare); Rome (Ciampino).

Embarked aircraft: Agusta AB 212ASW (62); Agusta-Sikorsky ASH-3D/H (40).

Shore-based aircraft: naval: Agusta-Bell 47G (4); Agusta-Bell 204AS (20); Agusta-Bell 47J (2); air force: Agusta-Sikorsky S-61R/HH-3F Pelican (18) (15)*; Breguet Atlantic 1 (15); Canadair CL-215 (4).

Units: navy: 1st Sqn (Sea King); 2nd Sqn (AB 212ASW); 3rd Sqn (Sea King); 4th Sqn (AB 212ASW); 5th Sqn (AB 212ASW); air force: 15° Stormo (Pelican); 30° & 41° Stormos (Atlantic).

Recent operations: In support of Gulf peacekeeping operations, two *Maestrale* class frigates (AB 212ASW) were deployed in September 1987.

Typical deployments: Task groups operate in the Mediterranean; most helicopters are embarked for short periods, remaining basically shore-based; naval and maritime patrol aircraft operate with NATO

Three distinct variants of Sea King are operated by the Italian Navy, including the weather radar-equipped ASH-3D which is optimised for utility, search and rescue and vertical replenishment duties. *(Paul Beaver)*

nations (particularly the US Sixth Fleet) but rarely deploy outside the Mediterranean Sea.

Weapon systems: Principal ASW weapons are Whitehead A/244S and Honeywell Mk 46 Mod 1 lightweight torpedoes; Sea Kings carry up to four and AS 212ASWs, up to two; Atlantics have internal capacity for eight torpedoes. Anti-shipping weapons are primarily the OTO Melara Marte Mk 2 guided missile which has replaced the Aerospatiale AS 12 for both helicopter types; the Atlantic is credited with the AS 12 and 127 mm unguided rockets but this cannot be confirmed. Primary ASW helicopter sensor is the Bendix AQS-13A/F dipping sonar family and sonobuoys can be deployed. There are reports that some of the Tornado IDS aircraft of the air force are armed with Kormoran 1 for anti-shipping duties and that the Mk 2 version will be purchased.

Personnel: naval: 45 000 (total); air force: 70 000 (total).

Remarks: The Italian Navy was confined to operating helicopters until 1987 when Parliament passed a law repealing the 1930s legislation which

Small ship helicopter operations continue to be flown by the Agusta-Bell 212ASW with its distinctive SMA Doppler radar above the cabin. It is also equipped with the Bendix AQS-13B/F dipping sonar. The helicopter is being dual-roled to take the Marte Mk2 anti-shipping missile. *(Paul Beaver)*

Italy's future long-range, medium ASW helicopter will be the EHI Merlin, a joint development between Agusta and Westland; it will enter service in 1992.

prevented the acquisition of fixed-wing aircraft. Although the air force Atlantics have been flown with naval systems operators and under naval operational control, the new light aircraft carrier, *Guiseppe Garibaldi*, was commissioned with helicopters only. It is thought that Italy will acquire a STOVL aircraft in 1989, the choice being between the Sea Harrier, AV-8B or Advanced Harrier II. Naval helicopter procurement is continuing with AB 212ASW and ASH-3D/H Sea Kings being delivered from Agusta and the air force has ordered a further 15 HH-3F Pelicans to keep the Brindisi production line open. Future plans rest on the EH 101, a joint Agusta-Westland project and the multi-national NH90. In the early 1990s, the Atlantics will need to be replaced.

Japan

Organisations: Maritime Self-Defence Force (MSDF); Maritime Safety Agency (MSA).

Organisational structure: Japanese naval aviation is an integral part of the Maritime Self-Defence Force and is organised into two commands: Fleet Air Force (all operational aircraft) and Air Training Command (training and support). the force headquarters is situated at Atsugi and flying and training control is from Shimofusa. Fleet Air Force is divided into six wings, each made up of a number of squadrons and attached to the local naval district. Air Training Command is divided into six squadrons which are allocated to four training groups around the country. The Maritime Safety Agency is very similar to the US Coast Guard in formation and would come under naval control in time of war; there are 200 ships at sea, a dozen of which carry helicopters for patrol and SAR operations. In addition, the MSA operates a number of fixed-wing, shore-based aircraft.

Command structure: MSDF: no details available.

Air-capable ships: MSDF: 1 'Aegis' type cruiser (building); 2 *Shirane* class destroyers; 2 *Haruna* class destroyers; 1 *Asagiri* class destroyers (7 building); 3 *Hatakaze* class destroyers (building); 12 *Hatsuyuki* class destroyers; 4 *Takatsuki* class destroyers; 1 large minelayer; 1 minelayer; 1 *Chiyoda* class submarine rescue ship; 1 *Sagami* class replenishment ship; 1 *Katori* class training ship; 1 icebreaker; MSA: 2 *Mitzuho* class cutters (building); 6 *Tsugaru* class cutters; 1 *Soya* class cutter.

Shore bases: MSDF: Atsugi; Hachinoe; Iwakuni; Kanoya; Komatsujima; Ohmura; Ominato; Ozuki;

Lockheed Orion aircraft are built under licence by Kawasaki Heavy Industries. The P-3C Orion is used for long range maritime patrol. *(Lockheed)*

Mitsubishi-built SH-3/HSS-2A Sea King helicopters will remain in service for anti-submarine duties well into the next decade and be replaced by the SH-60J Seahawk now under Japanese licenced development from Sikorsky.

Shimofusa; Tateyama; Tokushima; MSA: 13 in total, as MSDF and use of some civilian airports.

Embarked aircraft: MSDF: Mitsubishi/Sikorsky SH-3A/HSS-2A (50); Mitsubishi/Sikorsky HSS-2B Sea King (64); Mitsubishi (Sikorsky) SH-60J 12 (98)*; Sikorsky XSH-60J Seahawk (2). MSA: Bell 212 (29).

Shore-based aircraft: MSDF: Beech C-90 King Air (7) (2)*; Beech Queen Air (10); Fuji KM-2 (33) (2)*; Gates Learjet 36A (2)(1)*; Kawasaki KV 107-II-A-3 (AMCM) (7); Kawasaki EP-2J (2); Kawasaki P-2J (78); Kawasaki UP-2J Turbo-Neptune (2); Kawasaki/Hughes OH-6D (5); Kawasaki/Hughes OH-6J (3) (2)*; Kawasaki/Lockheed P-3C Update 2 Orion (75); Lockheed P-3C Orion Update 3 (12)*; Lockheed EP-3 (4)*; NAMC YS-11T (6); Shin Meiwa PS-1 (6); Shin Meiwa US-1 (10); (1)*; Sikorsky MH-53E Sea Dragon (4)*; Sikorsky S-62 (5); MSA: Beech Super King Air 200T (15); Bell 206B JetRanger (4); Lockheed C-130H-MP (1); Kawasaki/Hughes 369HS (2); NAMC YS-11A (2); Shorts Skyvan 3 (2).

Units: MSDF: (Fleet Air Wing 1) 1 Sqn (P-3C), 11 Sqn (HSS-2A/B), SAR Flt (S-61A); (Fleet Air Wing 2) 2 Sqn (P-3C), 4 Sqn (P-2J), SAR Flt (S-61A); (Fleet Air Wing 4) 3 Sqn, 6 Sqn (P-3C), SAR Flt (S-61A); (Fleet Air Wing 5) 5 Sqn (P-2J); (Fleet Air Wing 21) 101 Sqn (S-61A/HSS-2A/B), 121 Sqn (HSS-2A/B), 122 Sqn (HSS-2A/B); (Fleet Air Wing 31) 31 Sqn (PS-1), 71 Sqn (US-1), 81 Sqn (EP-2J/UP-2J/Learjet);

Mitsubishi-built SH-3/HSS-2A Sea King helicopters will remain in service for anti-submarine duties well into the next decade and be replaced by the SH-60J Seahawk now under Japanese licenced development from Sikorsky.

51 Sqn (OH-6D/J/P-2J/P-3C/HSS-2B); 61 Sqn (Queen Air/YS-11M); 111 Sqn (KV-107); Komatsushima Sqn (HSS-2A); Ominato Sqn (HSS-2A); Omura Sqn (HSS-2A); Antarctic Res Flt (S-61A); (Ozuki Air Trng Wing) 201 Sqn (Fuji); 221 Sqn (no aircraft); SAR Flt (S-62); (Tokushima Air Trng Wing) 202 Sqn (King Air/Queen Air); SAR Flt (S-62); (Kanoya Air Trng Group) 203 Sqn (P-2J); 211 Sqn (OH-6D/HSS-2A); (Shimofusa Air Trng Group) 205 Sqn (P-2J/YS-11T).

Recent operations: MSDF tasks have included the search for wreckage from KAL 007 and the continued shadowing of the increased activity by the Soviet Pacific Fleet. MSA has no recent reported operation, other than supporting the MSDF during the KAL 007 operation. Japanese Air Self-Defence Force helicopters are often used for overwater SAR tasks in co-ordination with the other two agencies; the HH-60J has been selected to replace the KV 107-11.

Typical deployments: Long-range patrol aircraft cover the Sea of Japan, South China Sea and eastern Pacific Ocean, while the helicopters operate from warships engaged in coastal and oceanic surveillance; Japanese warships deploy around the Pacific and South-East Asia, with the occasional training cruise to Europe and North America. The Maritime Safety Agency is more likely to be deployed to the islands of the Japanese archipelago, although the recent purchase of a Hercules LRMP aircraft indicates wider ranging tasks.

Weapon systems: MSDF: anti-shipping weapons include the recent purchase of the AGM-84 Harpoon air-launched semi-cruise missile for the P-3C; there are no helicopter-launched anti-shipping missiles at present although a programme to examine the procurement of a lightweight system is underway. ASW weapons include the Mk 44 and Mk 46 lightweight torpedoes for P-2, P-3 and HSS-2 aircraft.

Personnel: MSDF aviation: 8000; MSA: not available.

Remarks: The MSDF plans to withdraw the remaining six PS-1 ASW flying boats by March 1990 but will keep the US-1 amphibian in service for SAR. Future LRMP duties will be performed by the P-3C Update 3 and at least 30 new aircraft will be procured; the initial buy is 11. The last P-2J is scheduled to have been paid off by 1994 and ultimate requirement for the P-3C could be as many as 100 airframes. First deployments of the SH-60J Seahawk are expected in 1990 and the first helicopter assembled in Japan flew in August 1987; however JMSDF officers have privately said that they expect only 100 Seahawks to be built and that the EH 101 is the medium term goal. The HH-60J Rescue Hawk is expected to replace the KV 107-II-A-3 and SH-3A, with three being ordered in 1987 at a cost of US$ 49.2 million for the JASDF. The Japanese Air Self-Defence Force has ordered a total of five Sikorsky UH-60J Black Hawk helicopters for SAR. There are still delays with the AMCM replacement programme, although the Sea Dragon was selected in 1985, it is not destined to enter service until 1990/91.

One of the Shin Meiwa US-1A amphibians of the JMSBF; used primarily for SAR duties.

Korea, North

Organisation: People's Army Air force.

Organisational structure: Exact details are not easy to come by but it is thought that the People's Army Air Force is modelled on Soviet lines.

Command structure: Exact details are not available.

Air-capable ships: None.

Shore bases: Nampo, Sunan, Wonsan.

Embarked aircraft: None.

Shore-based aircraft: Ilyushin Il-28 'Beagle' (60) Mil Mi-14 'Haze A' (6).

Units: No details known.

Recent operations: None reported.

Typical deployments: No exact details known.

Weapon systems: Anti-shipping strike: 127 mm rockets and 27 mm cannon; anti-submarine: aerial torpedoes and mines.

Personnel: 50 000 (total).

Remarks: Very little is known about the People's Army Air Force, other than its predominant anti-South Korean posture. It is understood that the vast majority of the 'Beagles' in service are the Chinese B-5 versions.

Korea, South

Organisations: ROK Navy, ROK Coast Guard.

Organisational structure: Deriving its transport almost totally from the US, but with an important domestic industry. All maritime operations are conducted by the naval air element which consists of two medium-range squadrons and several units flying helicopters. The Coast Guard operates a single squadron of unarmed helicopters for a primary SAR role.

Command structure: Exact details are not available.

Air-capable ships: 5 *Gearing* (FRAM 1) class destroyers; 2 *Gearing* (FRAM II) class destroyers; 2 *Allen M Sumner* class destroyers; 5 *Ulsan* class frigates.

Shore bases: Chinhae, Inchon, Kunsan, Sachon.

Embarked aircraft: navy: Aerospatiale SA 319B Alouette III (10); Bell 206B JetRanger (5); McDonnell Douglas 500M/ASW (10); coast guard: Hanjin/MD 500D (10).

Shore-based aircraft: Grumman S-2A/F Tracker (20).

Units: No details available.

Recent operations: None reported.

Typical deployments: The main thrust of South Korean operations are to safeguard the sea routes and coasts against the potential threat posed by North Korea.

Weapon systems: Helicopters: armed with Mk 44 and Mk 46 lightweight torpedoes; Trackers: armed with 127 mm rockets for anti-shipping operations with Mk 46 torpedoes and depth bombs against submarines.

Personnel: navy: 20 000 (total); coast guard: not available.

Remarks: At the Farnborough Air Show in September 1988, Westland Helicopters announced that it had received the expected order for 12 Super Navy Lynx helicopters for the surface search and anti-ship role. The helicopters will be equipped with Ferranti Sea Spray MkIII radar and British Aerospace Sea Skua anti-ship missiles. South Korea is understood to have a further requirement for six light shipborne ASW helicopters and six medium ASW helicopters, probably for land-based operations.

South Korea ordered the Sea Skua-equipped Westland Lynx in 1988, mainly to counter the perceived threat from North Korean fast patrol boats.

Kuwait

Organisations: Kuwait Air Force.

Organisational structure: The Kuwait Air Force is based on the traditional British squadron basis and all military aircraft are controlled under its auspices.

Command structure: No exact details are known.

Air-capable ships: None.

Shore bases: Al Ahmadi, Al Jahra, Kuwait City.

Embarked aircraft: None.

Shore-based aircraft: Aerospatiale AS 332F Super Puma (12); McDonnell Douglas A-4KU Skyhawk (28).

Units: No exact details are known.

Recent operations: None reported.

Typical deployments: The Iraq-Iran conflict gave a greater emphasis to the protection of national boundaries and installations. There was concern in Kuwait that the conflict would spread to Kuwait as, despite an historic border dispute with Iraq, Iran considered the country to be a supporter of Iraq in the conflict.

Weapon systems: AS 332F is armed with Aerospatiale AM39 Exocet; the Skyhawks are flown for oilfield surveillance, armed with iron bombs and a variety of weapons which have not been announced.

Personnel: 1500 (total).

Remarks: Kuwait has a requirement for up to ten anti-ship missile-carrying helicopters for a contract to be placed in 1989.

Libya

Organisations: Libyan Arab Jamahiriyah Air Force.

Organisational structure: The basic organisation of the air force is along Soviet lines, with each regiment sub-divided into three squadrons. Interceptor aircraft are reported to be under a separate command. Preoccupation is with the intermittent land war with Chad, although the Libyans have been involved in actions against the US Sixth Fleet.

Command structure: No details are available. There are reports that the Tu-22s are Soviet-manned.

Air-capable ships: None.

Shore bases: Benghazi, El Adam (Nasser), Tripoli (Okba ibn Nafa), Tubruz (Tobruk), Zawiyat al-Bayda-Labraq (El Beida).

Embarked aircraft: None.

Shore-based aircraft: Aerospatiale SA 316B Alouette III (12); Aerospatiale SA 321M Super Frelon (6); Mil Mi-14 'Haze A' (12); Sukhoi Su-20 'Fitter'; Tupolev Tu-22 (7).

Units: No exact details are available.

Recent operations: No aggressive overwater operations have been reported in the last five years.

Typical deployments: Regular patrols are operated over the Gulf of Sidrah.

Weapon systems: SA 321M probably carries one Aerospatiale AM39 Excocet.

Personnel: 8000 (total).

Remarks: The Libyan forces are difficult to monitor, especially since the Chadian adventures and the defeat of late 1987. Soviet equipment is supplementing the ageing Italian and French-built aircraft, although the helicopter force is maintaining its serviceability. A number of aircraft have been lost in combat with US naval aircraft.

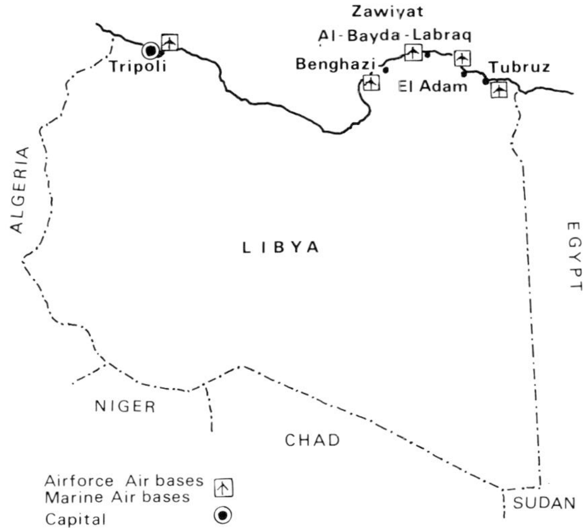

Malaysia

Organisations: Royal Malaysian Navy; Royal Malaysian Air Force.

Organisational structure: The Royal Malaysian Navy's new aviation branch is organised into a single squadron whereas the Royal Malaysian Air Force is organised into the Air Defence Command (combat aircraft) and Air Support Command (training and engineering).

Command structure: The basic operational unit is the squadron.

Air-capable ships: 2 Type 1500 frigates, 1 Yarrow frigate, 3 support ships, 1 survey ship.

Shore bases: navy: currently using air force facilities; air force: Kelaung (helicopter training base).

Embarked aircraft: navy: Westland Wasp HAS 1 (6).

Shore-based aircraft: air force: Aerospatiale SA 316B Alouette III (3); Lockheed C-130H-MP Hercules (3).

Units: No details available; air force: No 4 Sqn (Hercules), No 5 Sqn (Alouette).

Recent operations: None reported.

Typical deployments: navy: pilot and ground crew training began in 1987 with Australian assistance; air force: regular patrols of the Malacca Straits and surrounding sea areas. Major concerns are smuggling, illegal immigrants and piracy.

Weapon systems: navy: the Wasps will be armed with Mk 44 and Mk 46 lightweight torpedoes; air force: unarmed aircraft.

Personnel: navy: 10 000 (total); air force: 12 000 (total).

Remarks: The sale of six ex-UK Royal Navy Wasp HAS 1s was confirmed on 6 October 1987 and these helicopters represent the first naval aircraft for this country. Six further Wasps ordered for 1989.

The Westland Wasp is the first helicopter for the newly formed naval air arm in Malaysia; further equipment buys are expected by 1990. This cocooned helicopter was photographed prior to delivery. *(Westland)*

Malta

Organisations: Armed Forces of Malta.

Organisational structure: The armed forces are also called the Task Force and have a limited military role in this small island state.

Command structure: The only aircraft, helicopters, are flown by the Helicopter Flight which is organised along Italian lines.

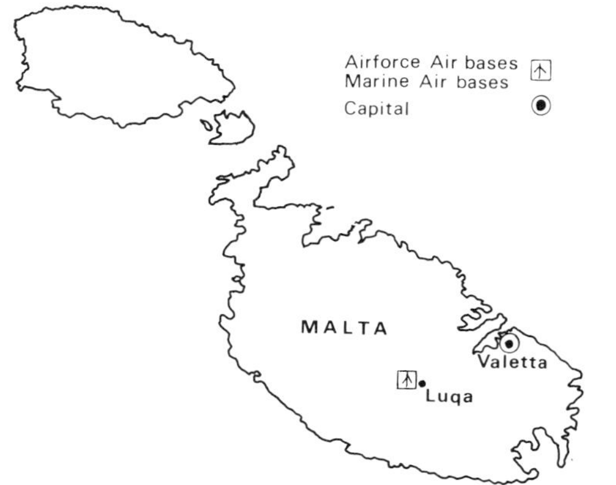

Air-capable ships: None.

Shore Bases: Luqa.

Embarked aircraft: None.

Shore-based aircraft: Agusta-Bell 204B (2); Agusta-Bell 206A (1).

Units: Helicopter Flight.

Recent operations: None reported.

Typical deployments: The AB 204B helicopters, loaned by the Italian government some years ago are used for coastal patrol and SAR duties around Malta and Gozo.

Weapon systems: None.

Personnel: 800 (total).

Remarks: A small force with a primary SAR role, mainly supported by the Italian Air Force, although there have been previous close relations with Libya.

Mauritania

Organisations: Mauritanian Islamic Air Force.

Organisational structure: The air force is organised into three squadrons of less than a dozen aircraft each. No combat aircraft are operated.

Command structure: No exact details are available.

Air-capable ships: None.

Shore bases: Nouadhibou, Nouakchott.

Embarked aircraft: None.

Shore-based aircraft: Piper Cheyenne II (2).

Units: Surveillance Sqn.

Recent operations: None reported.

Typical deployments: Coastal surveillance duties only.

Weapon systems: Unarmed.

Personnel: 200 (total).

Remarks: A small, limited force which is mainly preoccupied with supporting the country's infrastructure. However, maritime patrol duties are undertaken, especially to protect fishing rights. There has been no conflict over Spanish Sahara with the Polisario and Morocco since 1979.

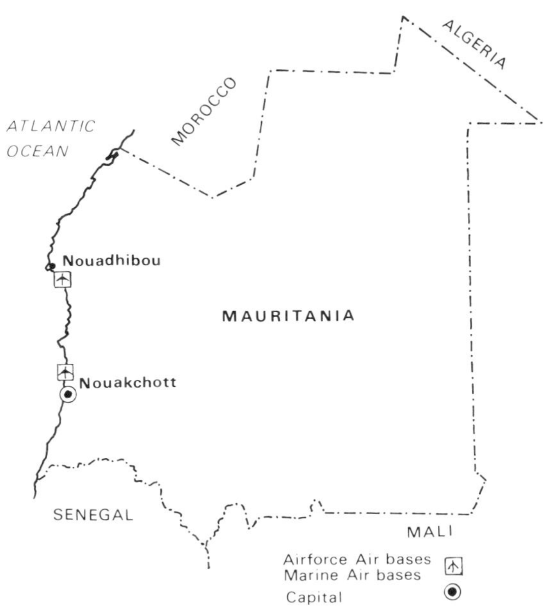

Mexico

Organisations: Mexican Naval Aviation.

Organisational structure: Naval Aviation has a maritime patrol and shipborne role to protect both coastlines. The coast guard force tasks some aircraft, especially the embarked BO 105CBS helicopters.

Command structure: No details are available.

Air-capable ships: 6 *Halcon* class frigates.

Shore-bases: Cozumel, El Cipres, Mazatlan, Mexico City, Pie de la Cuesta, Tampico, Vera Cruz.

Embarked aircraft: MBB BO 105C (6) (6)*.

Shore-based aircraft: Aerospatiale Alouette III (4); Bell 47G-3 (5); CASA C 212 Aviocar (6) (4)*; Cessna 337G (2); Cessna 402 (1); de Havilland Canada Buffalo (1); Gates Learjet 24 (2); Grumman HU-16A Albatross (8); Fairchild FH-227 (2); IAI Arava-201 (8); Mitsubishi MU-2 (1); training: Aerospatiale Alouette II (4); Beech T-34A Mentor (4); Beech Baron (2); Beech Bonanza (4); Cessna 150J (2); Cessna 182 (3).

Units: Patrol Sqn (HU-16); Reconnaissance sqn (helicopters); Transport Sqn Naval Aviation School.

Recent operations: None reported.

Typical deployments: The bulk of the shore-based operations are flown for coastal patrol, exclusive economic zone protection and fishery protection.

Helicopters are operated from a number of ships on coast guard duties.

Weapon systems: Unarmed, except for the HU-16 aircraft which have basic depth bombs.

Personnel: Not available.

Remarks: A funding shortage has prevented the Aviocars taking over the HU-16s role fully and no role equipment packages have been completed.

The sole shipborne helicopter type is the MBB Bo 105CBS for coastguard and SAR duties.

Morocco

Organisations: Royal Moroccan Air Force.

Organisational structure: The basic operational unit is the squadron of 10 to 12 aircraft. The air force is the only operator of military aircraft.

Command structure: Exact details are not available.

Air-capable ships: 3 *Champlin* class amphibious warfare ships.

Shore bases: Kednitra, Rabat, Tangier, Tantan, Tetouan.

Embarked aircraft: Agusta-Bell 206B (5).

Shore-based aircraft: Dornier Do 28D Skyservant (10).

Units: No details available.

Recent operations: Almost all air activity is devoted to operations against the Polisario Front in the former Spanish Sahara territory.

Typical deployments: Coastal patrol and anti-insurgent operations, especially in the south.

Weapon systems: Aircraft are unarmed.

Personnel: 12 000 (total).

Remarks: The bulk of the present and immediate future re-equipment plans are devoted to sustaining the war against the Polisario, although there are apparently plans to acquire helicopter-capable frigates in due course. No decisions about the helicopters to be embarked have been made.

Netherlands

Organisations: Royal Netherlands Navy (MLD); Royal Netherlands Air Force (KLu).

Organisational structure: There are five squadrons of aircraft which all have dual operational and conversion unit duties. Aircraft are based in the Netherlands and overseas at Hato on Curacao in the Dutch Antilles there is a single KLu coastguard/maritime patrol force. Pilot and aircrew training is undertaken by the government flying school and the KLu.

Command structure: The MLD has a flag officer in command, with staff officers dealing with the administration of the fixed-wing and rotary-wing types. Each squadron, the basic administrative unit, is commanded by a Lieutenant Commander. The commanding officer of the KLu detachment in Curaçao is Lieutenant Colonel Jan Looisen.

Air-capable ships: 2 *Tromp* class destroyers; 8 *Karel Doorman* class frigates (building); 2 *Jacob van Heemskerck* class frigates; 10 *Kortenaer* class frigates; 2 *Van Speijk* class frigates; 1 *Poolster* class replenishment ship; 1 *Improved Poolster* class replenishment ship.

Shore bases: De Kooy; Valkenburg; Hato (Curaçao).

Embarked aircraft: Westland SH-14B Lynx (9); Westland SH-14C Lynx (8).

Shore-based aircraft: Lockheed P-3C Orion update II (13); Westland UH-14A Lynx (5); KLu: Fokker F27 Maritime (2).

Units: 2 Sqn (P-3C); 7 Sqn (Lynx SAR); 320 Sqn (P-3C); 321 Sqn (P-3C); 860 Sqn (Lynx); KLu 336 Sqn.

Recent operations: None reported.

One of two F27 Maritimes delivered to the Royal Netherlands Air Force for maritime patrol purposes in the Netherlands Antilles.

The MLD (Royal Netherlands Naval Air Service) operates the P-3C Orion for long-range patrol from Valkenburg and Iceland for North Sea and North Atlantic Ocean operations. *(RNethN via F J Bachofner)*

Typical deployments: 1986-88: individual P-3C aircraft have been detached to Keflavik (Iceland) for NATO training as part of the defence force for the Greenland-Iceland-UK barrier.

Weapon systems: P-3C armed with AGM-85 Harpoon anti-shipping missile as well as Mk 44 and Mk 46 lightweight torpedoes, and depth bombs. It is possible that the MLD has a nuclear depth bomb role. The Lynx helicopters carry the Honeywell Mk 46 Mod 2 torpedo and BAe Mk 11 Mod 2 depth bombs. Lynx sensors include Alcatel DUAV-4 dipping Sonar (SH-14B) and AQS-81 MAD (SH-14C); UH-14A is unarmed. The F27 Maritimes are unarmed but have been equipped with a Global GNS 500A VLF Omega long-range navigation system as

Standard shore-based SAR helicopter is the Westland UH-14A Lynx.

Streaming the Texas Instruments ASQ-81(V) magnetic anomaly detector, this Westland SH-14B Lynx is embarked in HrMs *Evertsen*, one of the two remaining *Van Speijk* class frigates.

back-up to the Litton LTN 72 Inertial Navigation System.

Personnel: navy 16 800 (total).

Remarks: The MLD has been operating aircraft since the Second World War, including two spells of aircraft carrier operations. Today, it concentrates on maritime patrol duties in the North Sea and Northern Atlantic under NATO auspices, as well as operating Lynx helicopters for anti-submarine warfare and surface search duties from its destroyer and frigate force. The Atlantic LRMP aircraft have been withdrawn from service after two serious accidents and placed in reserve, having been replaced by the Orion. P-3C overhaul is carried out by KLM Royal Dutch Airlines at Schiphol airport.

Future programmes

Lynx replacement: The Netherlands Ministry of Defence has indicated that it will replace the existing shipborne and land-based Lynx helicopters with a common design. Contenders include the NH90 (NFH90 variant) and the Sikorsky S-70B Seahawk. The EH Merlin is considered too large for the Dutch requirement. The new helicopter will be capable of carrying anti-shipping missiles, anti-radiation missiles and modern lightweight torpedoes. It is understood that the Marconi Sting Ray and the Honeywell Mk 50 Barracuda are in competition to replace the MLD's Mk 46 torpedo inventory.

New Zealand

Organisations: Royal New Zealand Air Force.

Organisational structure: The Wasp helicopters are flown by Transport Wing's Naval Support Unit of 3 Sqn and the P-3Bs by the Maritime Wing of the RNZAF.

Command structure: The Operations Group, based at Whenuapai, controls the operational flying of the RNZAF and is divided into three wings, each commanded by the Group Captain. The Wings are divided into one or more squadrons, each commanded by a Wing Commander. Aircrew training is undertaken by Flying Training Wing. The Director of Naval Aviation is Commander P J W Stevens RNZN, who also heads Project Amokura to replace the Westland Wasp. The Project Amokura Technical Adviser is Wing Commander A G P Dick RNZAF.

Air-capable ships: 2 *Leander* Batch 3 frigates; 1 *Leander* Batch 1 (Ikara) frigate; 1 *Leander* class frigate; 1 survey ship; 1 support ship (building).

Shore bases: Auckland (Hobsonville), Whenuapai.

Embarked aircraft: Westland Wasp HAS 1 (10).

Shore-based aircraft: Lockheed P-3K Orion (6).

Units: 3 Sqn (Wasp), 5 Sqn (P-3B).

Long-range maritime patrol, including anti-submarine warfare, is the role of the Royal New Zealand Air Force's small fleet of Lockheed P-3K Orion aircraft. *(NZ Defence)*

The four *Leander* class frigates are complemented with a single Westland Wasp HAS1 helicopter each; the Wasps are controlled by the navy but operated by the air force's Transport Wing. *(NZ Defence)*

Recent operations: None reported, except to support the New Zealand government's efforts to restore peace in Fiji in 1987.

Typical deployments: The long-range maritime patrol aircraft support national foreign policy and general policing duties in the South Pacific.

Weapon systems: New Zealand has a non-nuclear policy. P-3s carry Mk 46 lightweight torpedoes and coventional depth bombs. The Wasp is armed with the Honeywell Mk 44 lightweight torpedo.

Personnel: 4300 (total).

Remarks: In November 1987, it was announced that a competition was underway for the Orion update programme with would include better sonobuoy processing equipment. Possible contenders are GEC Avionics and Magnavox. In November 1987, two of the original Westland Wasp helicopters remained in squadron service and eight others were operational from ex-UK Royal Navy stocks; the Wasp will be paid off in 1992.

Wasp replacement: Project Amokura has been formed to begin the selection process, with interest focused on a joint buy of the S-70B Seahawk for commonality with Australia for the ANZAC frigate programme, the design for which will be decided in late 1988.

Nigeria

Organisations: Nigerian Navy; Nigerian Air Force.

Organisational structure: The air force has been the traditional operator of aircraft but the delivery of Lynx shipborne helicopters has led to the development of the naval air arm. The air force operates maritime patrol aircraft and SAR helicopters in support of the navy.

Command structure: The basic naval unit is the squadron, commanded by a Lieutenant Commander, with each flight led by a Lieutenant. The Naval Staff Officer responsible for helicopter operations is a Commodore. The Navytown helicopter base is commanded by a Commander. The air force fixed-wing and helicopter assets devoted to maritime operations are thought to be operated by three separate squadrons, each commanded by a Wing Commander.

Air-capable ships: 1 *Aradu* class frigate.

Shore bases: Navytown (Lagos); Port Harcourt.

Embarked aircraft: Westland Lynx Mk 89 (3).

Shore-based aircraft: Dornier Do 128-6MPA (6); Fokker F27 Maritime (4); MBB BO 105C (20).

Units: No details are available other than that the

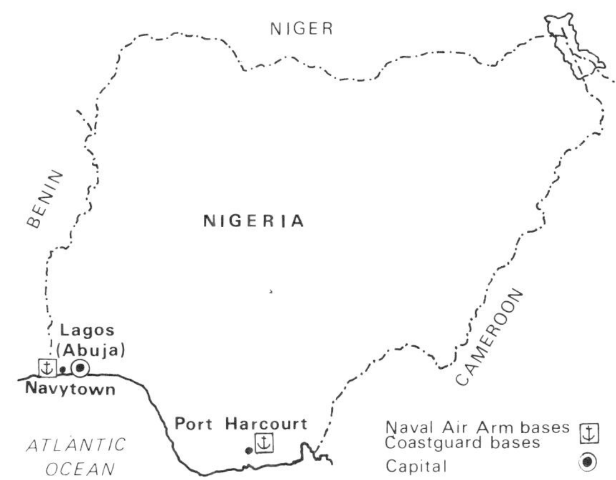

Lynx are operated by the Helicopter Squadron of the Nigerian Navy.

Recent operations: None reported.

Typical deployments: In late 1987, the first combined fleet exercises were undertaken and air force strike aircraft were shown as having an overwater role but it is not thought that they have a fully developed role yet.

Weapon systems: The Lynx helicopters are armed

Nigerian maritime patrol tasks are carried out by four Fokker F27 Maritime aircraft.

Two Lynx helicopters make up the embarked aviation element of the Nigerian Navy; they can be armed with the Whitehead A 244/S torpedo. *(Paul Beaver)*

with the Whitehead Motofides A/244 lightweight torpedo and can be fitted with 7.62mm GPMGs. Lynx primary is RCA 5000 radar.

Personnel: navy: 5000 (total), air force: 8000 (total).

Remarks: The Nigerian Navy is continuing to build up its force levels and may procure larger helicopters in due course; for the time being the air force retains the long range maritime patrol role.

Norway

Organisations: Royal Norwegian Air Force.

Organisational structure: The basic operational unit is the squadron which is organised on British lines with a Wing Commander in command, subordinate to the two geographically defined air commands.

Command structure: Northern Air command has its headquarters at Bodo and Southern Air Command at Bardufoss. The air force headquarters is at Oslo. The Lynx helicopters are operated by the air force on behalf of the Kystvakt (coast guard).

Air-capable ships: 1 logistic support ship, 3 *Nordkapp* class coastguard patrol ship.

Shore bases: Northern Air Command: Andoya; Banak; Bodo; Southern Air Command: Bardufoss; Orland; Rygge; Stavanger.

Embarked aircraft: Westland Lynx Mk 86 (5).

Shore-based aircraft: General Dynamics F-16A (56); Lockheed P-3B Orion (7); Lockheed P-3C Orion (4)*; Westland Sea King Mk 43/43A (8).

Units: Northern Air Command: 330 (Sea King); 331 (F-16); 333 (Orion); Southern Air Command: 330 (Sea King); 332 (F-16); 334 (F-16); 336 (F-16); 337 (Lynx); 338 Sqns (F-16).

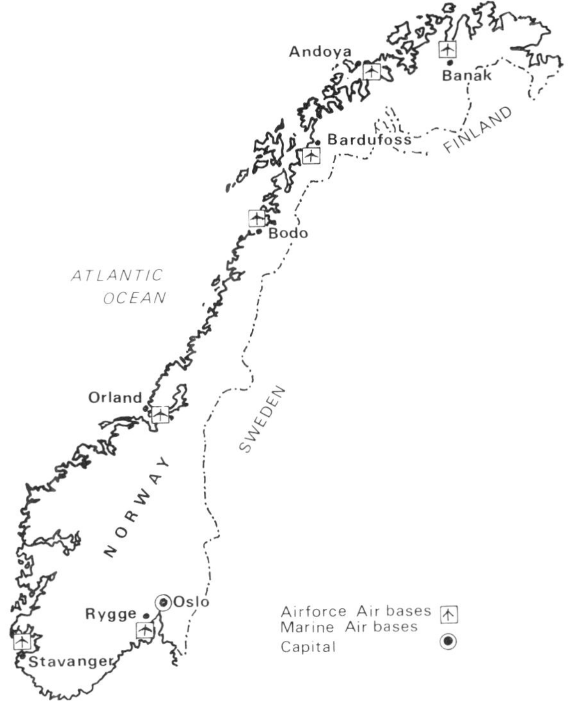

Norway's helicopter search and rescue force is made up of Westland Sea King Mk43/43As which provide cover for military and civilian operations. *(Westland)*

Shipborne operations are carried out for the Norwegian Coast Guard service by frigate-detached Westland Lynx Mk86 helicopters. *(Westland)*

Recent operations: None reported.

Typical deployments: The Orions keep a continuous watch on Soviet maritime activities around the North Cape, especially from the Kola Peninsula. In 1987, an Orion was accidentally rammed by a Soviet Air Force (VVS) Su-27 'Flanker' fighter during one of these patrols. The Lynx are used for fishery protection and the Sea King helicopters for medium-range SAR and in wartime, for combat rescue. The F-16 force provides anti-shipping strike capability.

Weapon systems: The major anti-shipping weapon is the NFT Penguin Mk 3 missile, locally developed from the Martin AGM-12C Bullpup. The Penguin is replacing the Bullpup on the anti-shipping strike F-16 fighters and has a range of over 20 mm (37 km) and carries a 50 kg warhead. The F-16 maintains a self-defence capability with the AIM-9L Sidewinder missile. It is possible that Norwegian P-3C Orion LRMP aircraft have the McDonnell Douglas Harpoon which gives the aircraft the extended anti-shipping capability. Standard ASW weapons are the Honeywell Mk 46 lightweight torpedo, carried by the Orions and presumably by the Sea Kings.

Personnel: 10 000 (total).

Norwegian waters are patrolled by the Lockheed P-3C Orion, operated by Northern Air Command.

Norway provides a striking force of 60 F-16A Fighting Falcons for overwater combat air patrols and anti-shipping strike roles.

Remarks: Norway plays a vital role in the protection of the NATO North Atlantic sea routes, against the potential of Soviet maritime forces. UK, Dutch and US forces would reinforce the Northern Flank of NATO (Norway) in the event of tension or conflict, identifying just how critical the safety of the northern air and naval bases are to the Alliance.

Oman

Organisations: Sultan of Oman's Air Force.

Organisational structure: The SOAF is the only operator of military aircraft and although there is limited maritime patrol (secondary duties of the Skyvan squadron) and no embarked aviation, uses the decks of the three air-capable ships. The basic operational unit is the squadron.

Command structure: The basic command structure follows British lines and is very often staffed with British-contract officers.

Air-capable ships: 1 Brooke Marine landing ship, 1 training ship, 1 logistic support ship.

Shore bases: Muscat, Salalah, Seeb.

Embarked aircraft: Aerospatiale AS 332C Super Puma (2), Agusta-Bell 206B JetRanger (3).

Shore-based aircraft: Agusta 205A (20); Shorts Skyvan 3M (15).

Units: No 2 Sqn (Skyvan); No 3 Sqn (AB 205); No 14 Sqn (AB 205 JetRanger); Royal Flight (AS 332C).

Recent operations: None reported.

Typical deployments: Coastal patrol and surveillance of shipping in the Gulf, Straits of Hormuz and Arabian Sea (Gulf of Oman). AB 205s have overwater SAR roll.

Weapon systems: None.

Personnel: 3000 (total).

Remarks: Oman was concerned about the protracted Iran-Iraq war which had threatened to widen and bring in Saudi Arabia and other Arab Gulf States. Following the 1988 ceasefire, a force improvement programme has been cut down because of the financial strain put upon the country's economy by the oil market decline, although funding for the air defence aircraft purchase continues. The Bell 214 ST fleet was re-deployed to army support in mid 1988 and replaced in the SAR roll by the AB 205, now equipped with flotation gear.

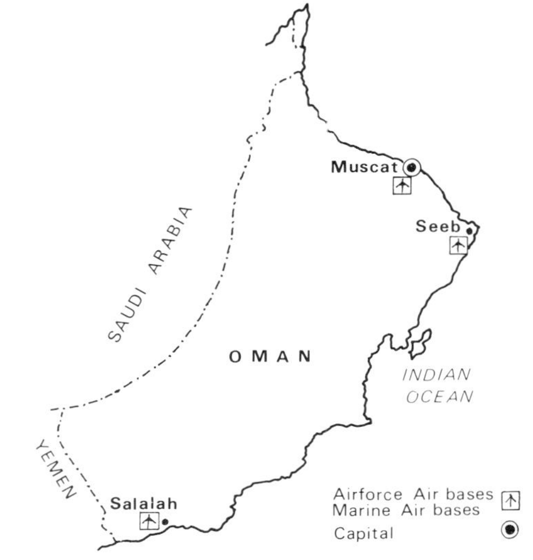

Agusta AB 205A single-engined SAR helicopters replaced Bell 214STs in 1988; the latter moving to onshore trooping tasks.

Pakistan

Organisations: Pakistan Naval Aviation.

Organisational structure: Flying training for the naval aviation force is provided by the air force.

Command structure: The naval aviation force is subordinate to the naval command, following an agreement with the air force for all maritime aviation to become a naval responsibility in 1975.

Air-capable ships: 1 'County' class destroyer (*Babur*).

Shore bases: Karachi, Korangi Creek, Mauripur.

Embarked aircraft: Aerospatiale SA 319B Alouette III (4).

Shore-based aircraft: Breguet Atlantic (3); Cessna 182 (2); Fokker F27 MPA Friendship (2); Westland Sea King Mk 45 (6).

Units: No details available.

Recent operations: None reported.

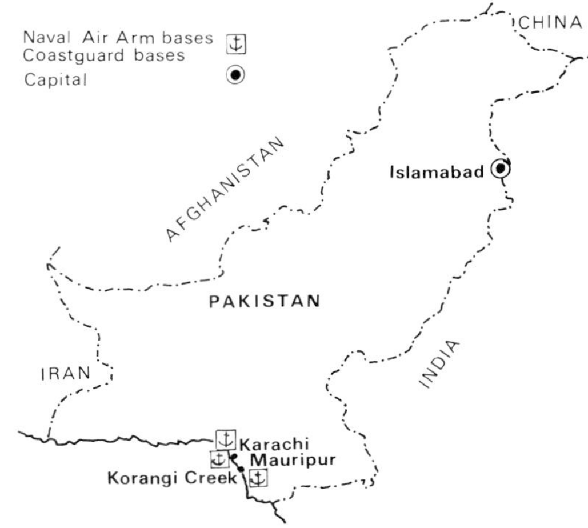

Typical deployments: It was reported in late 1986 that a detachment of USN Lockheed P-3C Orion aircraft have been deployed to Pakistan for Arabian Sea and Gulf surveillance duties. The aircraft are under US control. The Sea King and Alouette III helicopters are used for patrol, coastal surveillance and SAR duties, although the Sea Kings have a wartime role of anti-shipping strikes as well as anti-submarine warfare.

Weapon systems: The Atlantics have lightweight torpedoes and depth bombs, whilst the Sea King can be armed with the Aerospatiale AM39 Exocet missile, as well as lightweight torpedoes. The most likely source for torpedoes is Honeywell's Mk 46, although a close connection with the People's Republic of China is becoming very important.

Personnel: navy: 11 200 (total).

Remarks: In May 1987, Pakistan formally requested procurement of the Boeing E-3A Sentry (AWACS) aircraft but it seems likely that the Pakistan Air Force will receive the Grumman E-2C Hawkeye instead. In August 1988 the Pakistan government counter signed a letter of offer and acceptance to procure three Lockheed P-3C Orion maritime patrol aircraft manufactured by Lockheed Aeronautical Systems Company, manufactured USA which are scheduled to be delivered in 1991. The P-3C Orions are being purchased in a government-to-government transaction under the terms of a Foreign Military Sales agreement with the US Navy.

Although Pakistan has announced that it is buying the P-3C Orion under US Foreign Military Sales funding, the Fokker F27 Maritime continues to be an important medium-range overwater surveillance aircraft.

Panama

Organisations: Panama Air Force.

Organisational structure: No combat aircraft are operated because the country's defence is guaranteed by the United States.

Command structure: No details are available.

Air-capable ships: None.

Shore bases: Panama City.

Embarked aircraft: None.

Shore-based aircraft: CASA C 212 Aviocar (3); de Havilland Canada DHC-6 Twin Otter (2); Pilatus Britten-Norman Islander (2).

Units: No details available.

Recent operations: None reported.

Typical deployments: Coastal surveillance.

Weapon systems: Aircraft unarmed.

Personnel: 9500 (total including National Guard).

Remarks: A limited operational force with a limited operational role; this will change when the US hands back the Panama Canal Zone in 1999.

Papua New Guinea

Organisations: PNG Defence Force.

Organisational structure: A single squadron, Australian officered, provides limited maritime surveillance and all transportation duties.

Command structure: No details are available.

Air-capable ships: None.

Shore bases: Lae, Port Moresby, Rabaul.

Embarked aircraft: None.

Shore-based aircraft: GAF N22B Missionmaster (6).

Units: Air Transport Sqn.

Recent operations: None reported.

Typical deployments: Coastal surveillance and limited SAR only.

Weapon systems: Aircraft unarmed.

Personnel: 3350 (total Defence Force).

Remarks: Australia is the largest aid supplier and guarantees the country's independence, especially against the potential threat of Indonesian invasion.

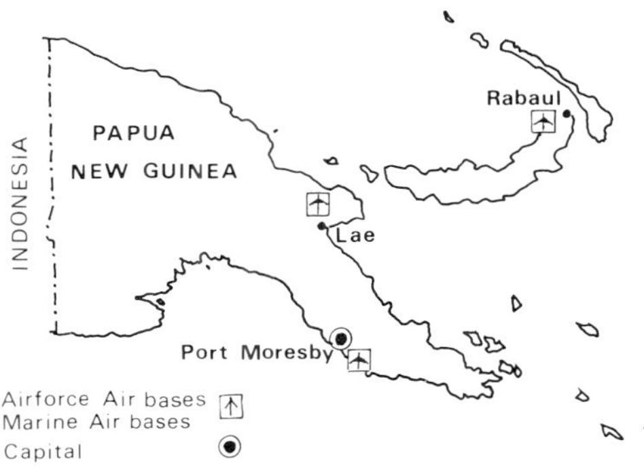

Paraguay

Organisations: Paraguayan Naval Air Service.

Organisational structure: Patrol and communication functions are carried out by the air service, although exact details are not available. Training is carried out in the Cessna 150Ms.

Command structure: Exact details not available.

Air-capable ships: 1 headquarters ship (platform only).

Shore bases: Asuncion (Campo Grande).

Embarked aircraft: None.

Shore-based aircraft: Bell 47G (2); Cessna 150M (4); Cessna U206A (4); Cessna 210 Centurion (1); Helibras HB 350B Esquilo (2); North American AT-6 Texan (2).

Units: No details available.

Recent operations: None reported.

Typical deployments: The air service is charged with keeping the country's river access open.

Weapon systems: Aircraft unarmed, although the Texans may retain 12.7 mm machine guns and the ability to carry 51 mm rocket projectiles.

Personnel: No data available.

Remarks: A small force used for river patrol and communications. It is independent from the air force. Re-equipment plans remain underfunded.

Peru

Organisations: Peruvian Naval Air Service.

Organisational structure: naval air service: there are a number of squadrons for maritime patrol, transportation and training, although some pilots attend the air force's academy.

Command structure: No details are available.

Air-capable ships: 1 guided missile cruiser (*Aguirre*), 4 *Lupo* class frigates.

Shore bases: Iquitos, Lima.

Embarked aircraft: Aerospatiale SA 316B Alouette III(3); Agusta-Bell 212ASW(6); Agusta-Sikorsky ASH-3D Sea-King (8).

Shore-based aircraft: Beech T-34A Mentor (2); Beech T-34C Turbo-Mentor (6); Beech Super King Air 200T (6); Bell 47G 96; Bell 206B JetRanger (8); Bell UH-1D (5); Douglas C-047 Dakota (5); Fokker F27 Maritime (1); Grumman S-2E Tracker (7): Grumman S-2G Tracker (4).

Units: No details available.

Recent operations: None reported.

Typical deployments: Limited scale operations around own coast and in support of counter-insurgency groups.

Weapon systems: The Trackers have both ASW and ASVW roles with lightweight torpedoes and rockets respectively; the Sea King and AB 212ASW helicopters carry the Whitehead Motofides A/244S lightweight torpedo; for shipborne anti-shipping operations, it is reported that some of the Sea King force is capable of being armed with the Exocet and that the 212s have a Marte Mk 2 capability.

Personnel: No details available.

Remarks: The Peruvian Navy has a requirement for a four-engined transport aircraft for longer range maritime patrol and long-range transport/passenger operations; it was reported in November 1987 that the Government had asked the US Air Force for two redundant C-130A Hercules for this task.

Aramada Peruana has operated a single Fokker F27 Maritime for more than ten years, supplemented by a number of Trackers and helicopters.

Philippines

Organisations: Philippines air Force, Philippine Naval Aviation.

Organisational structure: air force: the basic operational unit is the squadron, several of which make up a wing, depending on the aircraft type; naval aviation; a small force of fixed-wing aircraft and helicopters are thought to be organised as a single unit.

Command structure: air force: the Fokker F27 LRMP aircraft are operated by 15th Strike Wing; naval aviation: no exact details available.

Air-capable ships: 2 *Rizal* class corvettes.

Shore bases: air force: Sangley Point; naval aviation, Sangley Point.

Embarked aircraft: air force: none; naval aviation: PADC (MBB) BO 105C (5).

Shore-based aircraft: air force: Fokker F27 Maritime (3); naval aviation: PADC (PBN) Islander (9).

Units: air force: 27th Sqn (F27MPA); naval aviation: no details available.

Recent operations: None reported.

Typical deployments: both: coastal surveillance and SAR duties only.

Weapon systems: Aircraft unarmed.

Personnel: air force: 16 000 (total); navy: 26 000 (total).

Remarks: The new Aquino goverment has not considered the development or re-equiping of maritime aviation and it is thought that the country's budgetary situation will prevent any moves until 1990.

Fokker F27 Maritime of the Philippines Air Force, operated for anti-smuggler patrols throughout the archipelago.

Poland

Organisations: Polish Naval Aviation.

Organisational structure: The naval aviation force is organised into regiments of several squadrons each, along Soviet lines. Training is carried out by the air force, including that of naval personnel to act as ground crew.

Command structure: No details available.

Air-capable ships: None.

Shore bases: Elbag, Gdansk, Koszalin, Slupsk, Szczecin.

Embarked aircraft: None.

Shore-based aircraft: Ilyushin Il-28 'Beagle' (10); Mikoyan MiG-17F/LiM-6 'Fresco' (39); Mikoyan MiG-19 'Farmer' (40); Mil Mi-2 'Hoplite' (10); Mil Mi-4 'Hound A' (20); Mil Mi-8 'Hip-C' (8); Mil Mi-14 'Haze A' (15); Mil Mi-14 'Haze C' (4).

Units: No details available.

Recent operations: None reported.

Typical deployments: Support to naval forces, including Soviet Red Banner Baltic Fleet; surveillance of NATO and Swedish Baltic Sea operations, Reconnaissance carried by MiG-17 and MiG-19 aircraft.

Weapon systems: Fixed-wing aircraft carry a wide variety of weapons, including cannon, bombs and rockets for anti-shipping and amphibious assault operations. The 'Beagle' reconnaissance and anti-shipping unit could be armed with aerial torpedoes. No nuclear weapons are thought to be under Polish control. Mi-14 helicopters are armed with lightweight torpedoes, depth bombs and a variety of sonar equipment, including dipper sonar. The Mi-4 and Mi-8 helicopters are used for naval infantry assault duties.

Personnel: Not available.

Remarks: New Soviet equipment for air combat, reconnaissance and anti-shipping operations is thought to be forthcoming by 1990.

Demonstrating its amphibious capability, this Polish Navy Mi-14 'Haze C' is the last of a reported batch of 15 helicopters delivered in the early 1980s. Four of the batch are 'Haze C' SAR variants with the double door.

Portugal

Organisations: Portuguese Air Force (Forca Aerea Portuguesa).

Organisational structure: The Portuguese Air Force is directly subordinate to the country's government through an organisational system (CEMFA) which gives political control to the elected government via the Defence Minister. Air operations are centred at six air bases, with the operations group as the main administrative unit responsible for co-ordination of air activity within the air force. Each group has two or more squadrons reporting to it.

Command structure: Overall command is placed in the hands of the Chief of Staff, Portuguese Air Force and issued through the High Command (CEMFA). The basic operational unit is the squadron (esquadra). All squadrons numbered from 600 are maritime patrol but other units also fly MR sorties.

Air-capable ships: 3 *Vasco da Gama* class frigates (building); 4 *Baptista de Andrade* class frigates.

Shore bases: Lajes, Montijo.

Embarked aircraft: None.

Shore-based aircraft: Aerospatiale SA 330C Puma (12); CASA C 212A Aviocar (6); Fiat G91R (25); Lockheed C-130H Hercules (5); Lockheed P-3P Orion (6).

Units: Esq 303 (G 91R); Esq 501 (C-130H); Esq 502 (C212A); Esq 503 (C 212A); Esq 601 (Lockheed P-3P); Esq 751 (Puma); Esq 752 (Puma).

Recent operations: Portugal operates its naval aircraft in support of national and NATO requirements, apart from the Azores and exercises, such as Opengate, the aircraft are confined to Portuguese air space.

Typical deployments: The only non-mainland air base is at Lajes (Azores) where Puma helicopters are used for SAR and some of the remaining Fiat G 91Rs are flown for maritime reconnaissance and patrol duties. Lajes is of major strategic importance to NATO.

Weapon systems: Fiat G 91Rs are armed with Raytheon AIM-9 Sidewinder; Lockheed P-3P have a full suite of ASW weapons including lightweight Mk 46 Mod 2 torpedoes, depth bombs and the possibility of dropping mines.

Personnel: 9000 (total).

Remarks: Most aircraft overhaul is carried out by OGMA (Alverca do Ribatejo, north of Lisbon). P-3P Orion aircraft provided from NATO funds from RAAF contract after modernisation of Lockheed-California. The Portuguese Navy's bid for shipborne ASW helicopters for the MEKO class frigates should come to fruition in 1989 when the joint service Chief of Staff's appointment passes from the air force to the navy. The Westland Super Navy Lynx is the clear favourite in Portugal.

Aerospatiale SA 330C/L Pumas provide search and rescue facilities for mainland Portugal, Madeira and the Azores (752 Sqn).

Qatar

Organisations: Qatar Emiri Air Force.

Organisational structure: The Qatar Emiri Air Force is based on the traditional British squadron basis and all military aircraft are controlled under its auspices. The QEAF is commanded by Colonel Sheikh Hamad Bin Abdallah Al-Thani who is subordinate to the Commander-in-Chief, Major General Sheikh Hamad Bin Khalifa Al-Thani.

Command structure: The QEAF's combat aircraft are organised into four squadrons. The anti-surface vessel unit, flying the Commando has the primary responsibility for coastal surveillance and anti-shipping strikes.

Air-capable ships: None.

Shore bases: Doha International Airport.

Embarked aircraft: None.

Shore-based aircraft: Westland Commando Mk3 (8).

Units: 8 Sqn (anti-surface vessel).

Recent operations: Naval and air forces were involved in the border dispute with Bahrain over the Faisal al-Dibal reef in April–May 1986.

Typical deployments: The Iraq-Iran conflict gave emphasis to internal security, the protection of national boundaries and installations.

Weapon systems: Two Commando helicopters are each armed with a single Aerospatiale AM39 Exocet anti-shipping missiles; the rest with door-mounted 7.62 mm general purpose machine guns.

Personnel: 500 military, 300 civilian contractors (total).

Remarks: Qatar is a member of the Gulf Co-Operation Council and although is primarily responsible for its own coastline, it operates with other member states. The country has moved from the UK as a primary arms supplier and is now tied to a treaty with France signed in June 1987. The country was strictly neutral during the Iran-Iraq war.

Romania

Organisations: Romanian Air Force.

Organisational structure: The air force is the only operator of military aircraft, although the navy controls the operational use of helicopters patrolling the lower Danube river.

Command structure: No details available, although it is thought to follow standard Soviet lines.

Air-capable ships: 1 *Munteniá* class destroyer (building); 4 *Total* class frigates (platform only).

Shore bases: Constanta, Mamaia.

Embarked aircraft: IAR 316 Alouette III (6).

Shore-based aircraft: Mil Mi-4 'Hound' (6); Mil Mi-14 'Haze A' (6).

Units: No details available.

Recent operations: None reported.

Typical deployments: Support of River Danube and Black Sea naval operations, including those of the Soviet Black Sea Fleet; surveillance of NATO warships in the Black Sea.

Weapon systems: The helicopters are thought to carry lightweight torpedoes and depth bombs.

Personnel: 25,000 (total).

Remarks: It is thought that the 'Haze-A' helicopters have replaced the 'Hounds' for anti-submarine and surface search patrols. The recent adoption of a naval version of the locally licence-built IAR 316 Alouette III which possibly is equipped with nose-mounted search radar and hardpoints for ASW weapons is thought to be linked to the commissioning of *Munteniá* in 1985. Details are very difficult to obtain but it is thought that the second warship of the class will not be built thus making the expansion of the embarked naval air arm unlikely.

Saudi Arabia

Organisations: Royal Saudi Air Force; Royal Saudi Navy; Royal Saudi Coast Guard.

Organisational structure: air force: details of the organisation are not available but the squadron forms the basic operational unit; newly developed helicopter force only but operates the only embarked aviation. Coast Guard was formed in 1987/88 and no details are available.

Command structure: no details of any force are available.

Air-capable ships: 4 *Al Madinah* class frigates, 2 *Durance* class replenishment ships.

Shore bases: Al Qasuimah, Dhahran, Yanbuh.

Embarked aircraft: air force: none; navy: Aerospatiale SA 365F Dauphin 2 (20); coast guard: none.

Shore-based aircraft: air force: Boeing KE-3A Sentry (5); (8)*; AMD-BA Atlantique (ATL 2) (2*); navy: Aerospatiale SA 365F Dauphin 2; coast guard: Aerospatiale AS 332 F Super Puma (12)*.

Units: No details available.

Recent operations: None reported.

Typical deployments: Coastal protection and surveillance, especially in the Gulf because of the Iran/Iraq war.

Weapon systems: Dauphins are armed with the Aerospatiale AS 15TT guided missile for anti-shipping operations and could also carry lightweight torpedoes; Super Pumas will be armed with GIAT M621 20 mm cannon (six helicopters) and AM 39 Exocet (remaining six). Sentry aircraft are unarmed and the ATL 2 war load is awaiting confirmation of the programme.

Personnel: air force: 12 000 (total); navy: 2500 (total).

Remarks: Although the Boeing E-3A Sentry was delivered in 1981, there has been increasing concern over the RSAF's ability to operate the aircraft which during May 1987 still had US personnel aboard. The Dauphin helicopters were delivered under the Sawari programme with France, which included the joint development of the AS 15TT missile for which Saudi Arabia was the only user until 1989. The ATL 2 order still requires confirmation but shows an increased awareness of the maritime threat posed by the Iran/Iraq war; formation of Coast Guard and procurement of 12 Super Pumas confirms this awareness despite cessation of hostilities in September 1988.

SA 365F Dauphin launching AS 15TT.

Senegal

Organisation: Senegal Air Group.

Organisational structure: The air group is the only operator of military aircraft, but details of its organisation are not available.

Command structure: Details not available, but thought to be influenced by French practice.

Air-capable ships: None.

Shore bases: Dakar, Tambacounda, Zinguinchor.

Embarked aircraft: None.

Shore-based aircraft: de Havilland Canada DHC-6 Twin Otter (1); Embraer EMB-111 Bandeirante (1).

Units: No details available.

Recent operations: None reported.

Typical deployments: Coastal surveillance duties only; the development of the newly delivered OPV awaits a dedicated helicopter.

Weapon systems: The Bandeirante is armed for anti-shipping operations with 127 mm rockets.

Personnel: 500 (total).

Remarks: A continued development of this nation's defence force will continue with French and Brazilian assistance.

Seychelles

Organisations: Seychelles Defence Force.

Organisational structure: The small aviation component has been trained by Britain and India, following British lines.

Command structure: No details available.

Air-capable ships: None.

Shore bases: Mahe International Airport.

Embarked aircraft: None.

Shore-based aircraft: Fairchild Merlin IIIB (1); HAL (Aerospatiale) Chetak (Alouette III) (2); Pilatus Britten-Norman Maritime Defender (1); SOCATA Rallye 235 (2).

Units: No details available.

Recent operations: None reported.

Typical deployments: Transport and surveillance duties, as well as primary search & rescue.

Weapon systems: Aircraft thought to be unarmed.

Personnel: 1000 (total).

Remarks: The defence force also has strong links with Tanzania.

Singapore

Organisations: Singapore Air Force.

Organisational structure: Organised into 12 squadrons.

Command structure: The squadron is commanded by a Colonel or Lieutenant Colonel.

Air-capable ships: None.

Shore bases: Changi, Seletar, Sembawang.

Embarked aircraft: None.

Shore-based aircraft: Aerospatiale AS 322M Super Puma (22); Grumman EC-2C Hawkeye (4); Shorts Skyvan 3M (0).

Units: SAF: 125 Sqn (Super Puma); the Hawkeye unit has yet to be confirmed.

Recent operations: None reported.

Typical deployments: Super Puma ASW/SAR helicopters were declared operational in July 1986; Hawkeye and Skyvan aircraft fly Malacca Straits surveillance.

Weapon systems: No details available.

Personnel: 6000 (total).

Remarks: There are 12 further Super Pumas on option through SAMAERO/SAMCO.

One of four Grumman E-2C Hawkeye early warning and control aircraft for the Republic of Singapore Air Force. *(Dr Tan Peng Hui)*

Coastal surveillance and air-sea search and rescue duties are undertaken by six Shorts Skyvan 3M patrol aircraft. *(Shorts)*

Somalia

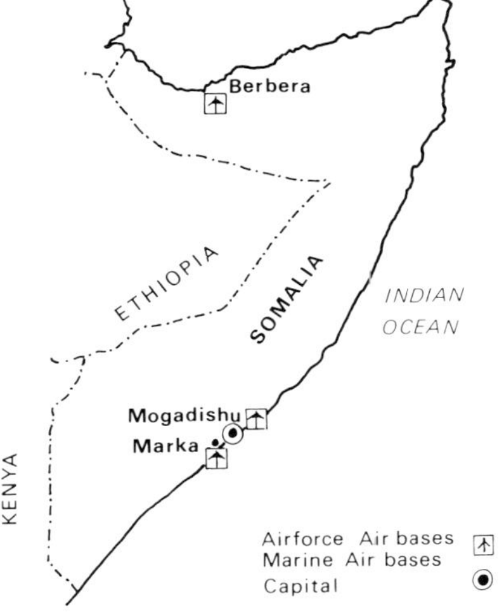

Organisations: Somalia Aeronautical Corps.

Organisational structure: Details are not available.

Command structure: Details are not available.

Air-capable ships: None.

Shore bases: Berbera, Marka, Mogadishu.

Embarked aircraft: None.

Shore-based aircraft: Agusta-Bell 204B (1).

Units: No details available.

Recent operations: None reported.

Typical deployments: Search and rescue only.

Weapon systems: Unarmed helicopter.

Personnel: No details available.

Remarks: A single helicopter is not effective as a maritime aviation force but it is unlikely that other assets will be made available because of the continued funding restrictions and the threat from Ethiopia.

South Africa

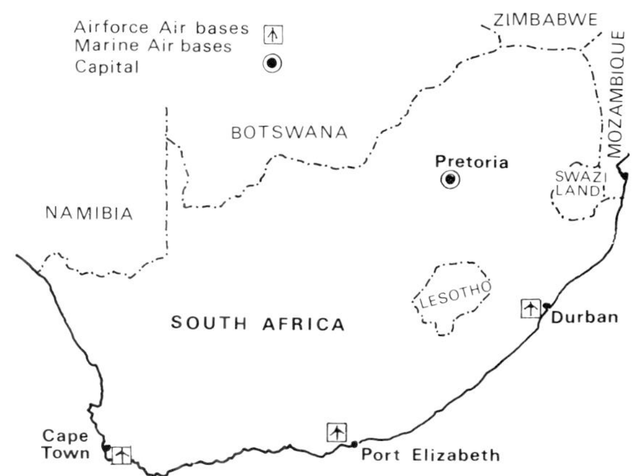

Organisations: South African Air Force.

Organisational structure: The air force operates aircraft on behalf of the other armed services, with Maritime Command tasked with providing the maritime reconnaissance and shipborne aircraft for national defence. Southern Air Command in the Cape provides Puma and Super Frelon helicopters for the replenishment ships *Tafelberg* and *Drakensberg*.

Command structure: The squadron is the basic unit, commanded by a Lieutenant Colonel and split into flights. The helicopter detachments for warships are commanded by a Major (or Captain) with a full supporting ground crew.

Air-capable ships: 1 *President Kruger* class frigate (1); 1 fleet replenishment ship (*Tafelberg*); 1 special auxiliary (*Drakensberg*); 1 *Hecla* class survey ship.

Shore bases: Cape Town (D F Malan), Durban, Port Elizabeth, Ysterplaat.

Embarked aircraft: Aerospatiale SA 330E/H/J Puma (10); Aerospatiale SA 321 L Super Frelon (2); Westland Wasp HAS 1 (9).

Shore-based aircraft: Douglas C-47 Dakota/Dakelton (20); Piaggio P166S Albatross (18).

Units: Maritime Command: No 22 (Alouette III/Wasp); No 25 Sqn (C-47); No 27 (Piaggio); No 35 Sqn (C-47); Southern Air Command: No 15 Sqn (Alouette III/Puma); No 30 Sqn (Puma); No 31 Sqn (Super Frelon/Alouette III).

Recent operations: It is understood that Maritime Command aircraft have supported air operations in Namibia and SAR has been carried out near Reunion (for the South African Airways Boeing 747SP) and in the Mozambique Channel.

Typical deployments: Surveillance of the Cape shipping routes, coastal protection and counter-insurgency are the main roles.

Weapon systems: The helicopters are armed with light machine guns and depth bombs of local manufacture; it is not known whether there are any aerial torpedoes in service. The C-47s have no armament.

Personnel: 10 000 (total).

Remarks: The continuation of the United Nations arms embargo means that all SAAF equipment is refurbishment or locally produced; no replacement for the Shackleton (phased out in 1984) seems likely.

The South African Air Force's Southern Command operates the SA 321 Super Frelon and the SA 330H Puma for embarked support aboard its fleet support ships. Both types have locally developed manual main rotor blade folding. *(SAAF photo)*

Spain

Organisations: Spanish Navy (Arma Aerea de la Armada); Spanish Air Force.

Organisational structure: navy: there are six naval aviation units, known as Esquadrilla, all based at Rota with detachments to ships at sea and other airfields as necessary; as all embarked aircraft are navy operated. Air force: the long range maritime patrol aircraft are organised under the command of Patrol Wing 22 (part of the Tactical Air Command/MATAC) and the Canaries Command (MACAN).

Command structure: navy: the basic administrative unit is the squadron (esquadrilla), commanded by a Lieutenant Commander; there is a flag officer for naval aviation; air force: the patrol wing is commanded by a Colonel and the MACAN squadron comes under the direct command of the general commanding the Canaries.

Air-capable ships: 1 aircraft carrier (*Dedalo*); 1 STOVL aircraft carrier (*Principe de Asturias*); 1 *Roger de Lauria* class (FRAM II) class destroyer; 5 *Gearing* (FRAM I) class destroyers; 4 *Oliver Hazard Perry* class frigates; 2 *Paul Revere* class amphibious warfare ships; 1 survey ship; 1 *Casado* class transport.

Shore bases: navy: Rota; air force: Jerez (Orion); Gando (F27MPA).

A McDonnell Douglas/British Aerospace EAV-8B Harrier II of the Spanish naval air arm is refuelled over the Atlantic Ocean on its delivery flight to Rota. Twelve EAV-8B are in service. *(McDonnell Douglas)*

McDonnell Douglas/British Aerospace AV-8S Matador STOVL aircraft remain in service for training tasks.

Embarked aircraft: Agusta AB; 212ASW (11); British Aerospace/McDonnell Douglas AV-8S Matador (8); McDonnell Douglas/Hughes 369M (11); McDonnell Douglas EAV-8B Matador II (12); Sikorsky SH-3D/G Sea King (10); Sikorsky SH-3D Sea King AEW (3); Sikorsky S-70L Seahawk (6)*.

Shore-based aircraft: : navy: Agusta-Bell 47G (4); Bell 47G (6); British Aerospace/McDonnell Douglas TAV-8S Matador (2); CASA C 212–200 MP Aviocar (7); Canadair CL-215 (13); Cessna Citation II (2); Piper Commanche (2); Piper Twin Commanche (2); air force: Aerospatiale AS 332 M Super Puma (10); Fokker F27 Maritime (3); Lockheed P-3A Orion (6); Lockheed P-3B Orion (5).

Units: Esquadrilla 001 (Bell 47G); Esq 003 (AB 212ASW); Esq 004 (light transports), Esq 005 (Sea King); Esq 006 (Hughes 369M); Esq 008 (AV-8S/ EAV-8B).

Recent operations: None reported.

Typical deployments: National and NATO operations in the Eastern Atlantic and Mediterranean Sea, including the protection of North African colonies and the Canary Islands.

Fleet fighter training is carried out in the McDonnell Douglas/British Aerospace TAV-8A tandem-seat trainer. *(Patrick Allen)*

Canadair CL-215 amphibians are used for coastal patrol and search and rescue duties by the Spanish Air Force. *(Canadair)*

Weapon systems: The naval helicopters are armed with the Honeywell Mk 46 Mod 1 & 2 lightweight torpedo for anti-submarine warfare (but Spain is still deciding on the ASW weapon for the S-70B); depth bombs are also carried. The Orion patrol aircraft have a full suite of ASW torpedoes and depth bombs; the F27MPA has similar weapons.

MBB BO 105CBs of the Spanish Customs Service. *(via Brian Walters)*

CASA 212 Aviocar of the Spanish Air Force and used for maritime patrol and search and rescue duties.

Personnel: navy: 32 000 (total); air force: 21 000 (total).

Remarks: Equipment plans include twelve McDonnell Douglas/British Aerospace. EAV-8B Harrier II which completed its maiden flight in August 1987 for first delivery in late December 1987. The aircraft will supplement the existing force of AV-8 Matadors. They will be embarked in the new STOVL carrier which will replace *Dedalo* and are part of the late 1980s re-equipment programme for the Armada. The first embarkation was due in late 1988 during *Prinicipe de Asturias* operational work-up. The naval air service will be well equipped by 1992 and the air force is seeking to bring its Orion force up to scratch by supplementing with ex-Royal Norwegian Air Force P-38s after modernisation by Lockheed.

Spanish Air Force helicopter SAR is provided by the AS 332 Super Puma.

Light shipborne ASW continues to be provided by the McDonnell Douglas 500MD/ASW, equipped with Bendix RDR-1300 search radar, a Texas Instruments ASQ-81 MAD and a single Mk44 torpedo.

Sudan

Organisations: Sudanese Air Force.

Organisational structure: The air force is the only operator of military aircraft but data is not available on the full organisation profile.

Command structure: No details known.

Air-capable ships: None.

Shore bases: Port Sudan.

Embarked aircraft: None.

Shore-based aircraft: CASA C 212 Aviocar (2).

Units: No details known.

Recent operations: None reported.

Typical deployments: Coastal surveillance in the Red Sea.

Weapon systems: Thought to be unarmed.

Personnel: 2500 (total).

Remarks: Not a strong force as funding is restricted.

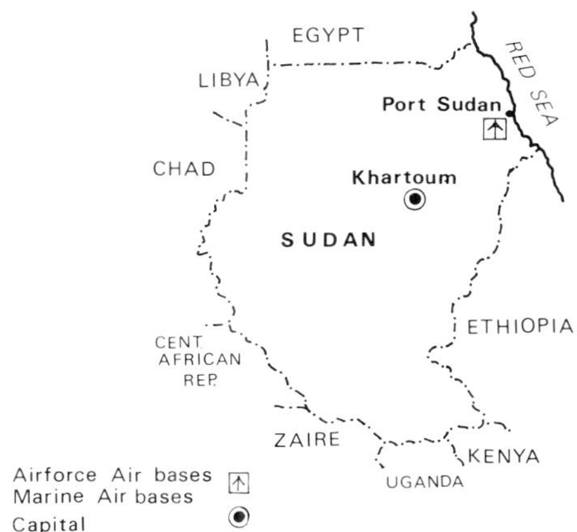

Sweden

Organisations: Royal Swedish Navy (RSwN); Swedish Air Force (SwAF).

Organisational structure: RSwN: There are three naval air squadrons, supported by training from the air force; SwAF: the air force is organised by wings on a type and geographical basis; maritime support is provided to the navy for anti-shipping reconnaissance and strike, as well as specialist EMC/ESM/ELINT/COMINT services. The air force provides all aircrew training.

Command structure: RSwN: the three units are each commanded by a Lieutenant Commander, subordinate to the appropriate flag officer at each base. There is no command formation; SwAF: each wing has three subordinate squadrons, distributed around the country's six military districts; each squadron is commanded by a Lieutenant Colonel. The senior officers in naval aviation are: Helicopter division Or1BO, Captain Lars Thomasson RSwN; Helicopter division MKV, Lieutenant Commander Einar Gjellan RSwN; Helicopter division Or1B S, Lieutenant Commander Roger Eliasson RSwN.

Air-capable ships: 1 *Carlskrona* class mine warfare ship; 2 *Alvsborg* class minelayers, 3 *Urho* class icebreakers, 1 *Oden* class icebreaker; 1 *Tor* class icebreaker.

Shore bases: RSwN: Berga, Kallinge, Save; SwAF: Lulea, Malmslatt, Norrkoping, Ronneby, Soderhamm.

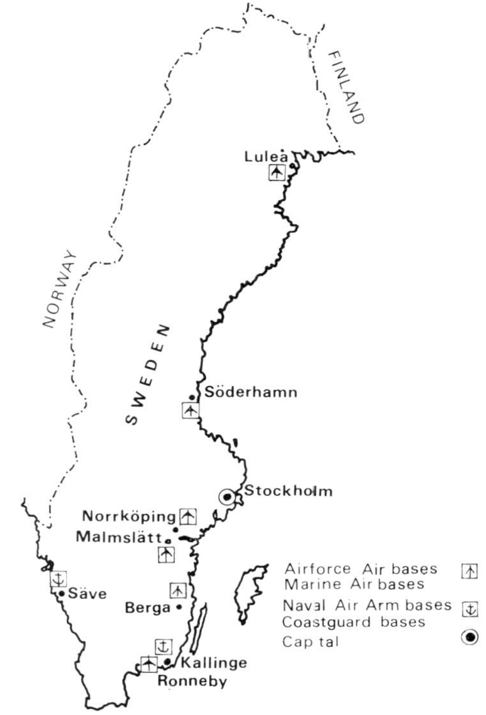

Royal Swedish Air Force long-range overwater strike operations are flown by the SAAB SH37 Viggen which can be armed with air-to-surface missiles.

Embarked aircraft: None although helicopters can operate from the air-capable ships.

Shore-based aircraft: RSwN: Aerospatiale Alouette II (12); Agusta-Bell 206A JetRanger/Hkp-6B (9); Boeing 107-II-5/Hkp-4B (3); Kawasaki (Boeing) KV 107-II/Hkp-4C (4); SwAF: Aerospatiale AS 332M1 Super Puma (10); Boeing 107-II/Hkp-4A (10); Saab-Scania SH 37 Viggen (24); Sud Aviation Caravelle III/Tp85 (2).

Units: RSwN: 1st Helo Sqn (Alouette II/JetRanger); 2nd Helo Sqn (KV 107); 3rd Helo Sqn (BV 107); SwAF: F13 Wing (Viggen), F13M Wing (Caravelle), F17 Wing (Viggen/BV 107); F21 Wing (Viggen).

Recent operations: RSwN: during the recently increased foreign submarine activity in Swedish territorial waters, more anti-submarine operations have been mounted using helicopters (some transferred from the air force); no successes were claimed; SwAF: none reported.

Typical deployments: RSwN: helicopters patrol the

The Boeing/Kawasaki KV107 tandem-rotor helicopter is flown by the SwAF (right) and the RSwN (left) for SAR and anti-submarine warfare respectively. Both variants are destined to remain in sevice for some years.

approaches to naval installations; SwAF: Viggen aircraft are used to identify surface ships approaching Swedish waters and the Caravelle carriers electronic surveillance. Super Puma helicopters operate long-range overwater SAR in Baltic Sea.

Weapon systems: The country manufactures most of its own weapons, including the FFV Type 42 lightweight torpedo, the Type 422 reduced charge type and the improved Type 431. Viggen aircraft carry the Saab RBS-15F anti-shipping missile with a range of at least 80 km.

Personnel: RSwN: 10 000 (total); SwAF: 9000 (total).

Remarks: The new air force SAR helicopter, a replacement for the existing Boeing 107 fleet, is the Super Puma which will be delivered between 1989 and 1990. Super Puma equipment includes the Racal Avionics RAMS 3000 management system. After the notorious 'Whiskey on the rocks' incident with a Soviet SSK (conventional submarine) grounding within Swedish territorial waters, increased use has been made of the Vertol 107 design, both Boeing and Kawasaki built, including the development of the Type 431 torpedo, new acoustic processing and self-defence countermeasures; the helicopters are also being re-engined with updated Rolls-Royce Gnome powerplants.

MBB BO 105CBs short-range SAR helicopter of the Royal Swedish Air Force.

Syria

Organisations: Syrian Arab Air Force.

Organisational structure: This air force is organised along Soviet lines and is the only operator of military aircraft.

Command structure: The basic unit is the squadron, three of which make up a regiment.

Air-capable ships: None.

Shore bases: Aleppo (Halab-Neirab), Hamah.

Embarked aircraft: None.

Shore-based aircraft: Kamov Ka-25 'Hormone' (3); Mil Mi-14 'Haze A' (14).

Units: No details available.

Recent operations: None reported.

Typical deployments: Coastal surveillance, support of national and Soviet operations in the Mediterranenan.

Weapon systems: Presumed to be Soviet-designed lightweight torpedoes and depth bombs.

Personnel: 70 000 (total).

Remarks: The 'Haze A' helicopters have recently replaced the 'Hormone A' types as a counter to the perceived increase in Israeli submarine activity, but it is surprising that such an important nation in the Middle East has no greater maritime presence. There are apparently no long-range maritime surveillance/strike aircraft in the inventory.

Taiwan

Organisations: Republic of China Air Force, Republic of China Navy, Republic of China Marine Corps.

Organisational structure: ROCAF: very much based on US lines with a large officer corps and general staff, naval aircraft are crewed by the air force. The Albatross amphibians are mainly used for SAR in support of 455 Tactical Fighter Wing; ROCN: largely based on US lines because of the equipment operated and the training given by the United States; ROCMC: follows the ROCN and is only a small force.

Command structure: ROCAF: based on the three squadrons, subordinate to the wing, with a Lieutenant Colonel commanding a squadron; ROCN: based on the squadron and the subordinate detachment (embarked flight); ROCMC: no details available.

Air-capable ships: 12 *Gearing* (FRAM II) class destroyers, 1 *Gearing* (FRAM II) class destroyers, 2 *Allen M Sumner* (FRAM II) class destroyers, 1 *Cabildo* class dock landing ship, 1 *Wu Kang* class transport.

Shore bases: ROCAF: Chiayi, Pingtung; ROCN & ROCMC: Chaiyi, Pingtung.

Embarked aircraft: McDonnell Douglas 500MD/ASW (10).

Shore-based aircraft: ROCAF: Grumman HU-16B Albatross (6); ROCN: Beech King Air 200 (1); Grumman S-2A/E Tracker (29); Grumman S-2F Tracker (9); Sikorsky S-58/HSS-1 (10); Sikorsky S-70C (10); ROCMC: McDonnell Douglas 500M (6).

Units: ROCAF: 4 Sqn (Albatross); ROCN & ROCMC: no details.

Recent operations: None reported.

Typical deployments: Coastal protection, with continued emphasis on the perceived threat from the People's Republic of China. It is understood that the S-70C search and rescue helicopters are based at Chiayi and the S-2E Trackers are co-located with the air transport wing's C-130 Hercules and C-119 Packet units.

Weapon systems: The MD 500MD ASW helicopters are armed with up to two Honeywell Mk 46 Mod 1 lightweight torpedo, whilst the Trackers are armed

Fixed-wing support for naval operations is supplied by 29 Grumman S-2A/E Tracker aircraft, based at Pingtung; in addition there are nine S-2F Trackers in service, modified with US funding some years ago. *(DTM)*

Most of the Taiwanese destroyers are air-capable and carry a single McDonnell Douglas 500MD/ASW helicopter embarked; this is *Dang Yang*, the single *Gearing* FRAM II destroyer. *(DTM)*

Taiwanese 500MD/ASW helicopters are equipped with the Bendix RDR-1300 search radar, Texas Instruments AN/ASQ-81 MAD and a Mk44 Mod3 torpedo. *(DTM)*

with a suite of torpedoes and depth bombs for ASW and 127 mm rockets for anti-shipping operations. An air-launched version of the Hsiung Feng II anti-shipping missile will be operational soon.

Personnel: ROCAF: 70 000 (total); ROCN: 35 000 (total); ROCMC: 29 000 (total).

Remarks: In 1986, it was announced that US FMS facilities would be made available for the Taiwanese Navy to refurbish the remaining Tracker airframes for a further 20 years life, including new turbine engines (plans for which were shown at the Asian Aerospace Show in 1988). P-3C Orion long-range maritime patrol aircraft are reportedly on order but funding and procurement is becoming more difficult. There are strong links between South Africa, Israel and Taiwan.

First illustration of Sikorsky S-70C Black Hawk.

Thailand

Organisations: Royal Thai Navy.

Organisational structure: The naval air arm has been upgraded in recent years for fixed-wing maritime patrol and helicopter operations. The Searchmasters are funded by the Australian government.

Command structure: Believed to follow typical US lines with the squadron as the basic unit; no further details available.

Air-capable ships: 2 *Jiang-Hu* class frigates (on order).

Shore bases: Satahip.

Embarked aircraft: Bell 212 (2).

Shore-based aircraft: Bell 205/UH-1H (4); Bell 212 (6); Bell 214ST (4); Canadair CL-215 (2); Cessna 337 (10); Fokker F27 Maritime (3); Fokker F27-400 (2); GAF Searchmaster B (5); Grumman S-2F Tracker (10).

Units: No details available.

Recent operations: None reported.

Typical deployments: F27s are used for target towing and SAR as well as for more traditional maritime patrol and EEZ enforcement roles. The Bell helicopters are used for transportation, coastal surveillance and reportedly, the 212s have an anti-submarine warfare role.

The Royal Thai Navy operates four Bell 214ST helicopters for transport, general support and some assault duties. One of the original batch was destroyed very soon after delivery and has now been replaced. *(Bell)*

Medium range overwater surveillance and patrol tasks are undertaken by the Royal Thai Navy's fleet of F27 Maritime Enforcer aircraft. Besides the Sting Ray air-launched torpedo and Mk11 depth bomb, the aircraft will be equipped for the Harpoon anti-ship missile. This has been made possible through a Foreign Military Sales contract between the US Navy and McDonnell Douglas Defense & Electronics for a $3.385 million integration package to be completed by December 1988 by Fokker in the Netherlands. *(Fokker)*

Weapon systems: Fokker has provided a full suite of ASW weapons for the F27MPA aircraft, including the Marconi Underwater Systems Sting Ray lightweight torpedo.

Personnel: 20 000 (total).

Remarks: First two Bell 214ST for SAR and liaison duties delivered in August 1987, but one crashed within days but it is presumed that it will be replaced. The two Grumman HU-16B Albatross were replaced by the Canadair CL-215. The Royal Thai Navy will convert two of its eight Bell 212 utility and support helicopters to ASW with the delivery of two helicopter capable Chinese-built frigates in 1992. The ASW package will be project managed by Bell Asia at Singapore and it represents the first ASW configuration for Bell-built (rather than Agusta-built) Model 212s. Thailand's fleet of Bell 214STs are now being used for VIP transport as Admiral's Barges. All Royal Thai Navy helicopters are capable of door-mounting pintle-type 7.62 mm M60 machine guns.

Turkey

Organisations: Turkish Naval Aviation.

Organisational structure: Originally formed in 1972 and still trained by the air force, the naval aviation arm is continuing to expand with an anti-shipping capability for the MEKO 200 frigates, using modified AB 212ASWs.

Command structure: Thought to be based on US lines, but no exact details available although it is known that the first AB 204 squadron formed was allocated to the Dardanelles Naval District. Naval aviation headquarters is thought to be at Karamursel.

Air-capable ships: 2 *Carpenter* (FRAM I) class destroyers; 2 MEKO 200 class frigates (2 more building); 2 *Berk* class frigates (platform only); 2 tank landing ships; 1 submarine tender; 1 *Dixie* class destroyer tender (platform only).

Shore bases: Izmir (Cigli); Izmir (Cigli-West); Izmir (Cumaovasi); Izmir (Gaziemir); Karamursel; Kutahya.

Embarked aircraft: Agusta AB 212ASW (12).

Shore-based aircraft: Agusta AB 204AS (3); Agusta AB 205A (3); Grumman S-2A Tracker (8); Grumman S-2E Tracker (15); Grumman S-2E Update (18); Grumman TS-2A Tracker (2).

Units: No details available.

Recent operations: None reported.

Typical deployments: Patrol and surveillance operations in the Baltic Sea (as part of the NATO Soviet watch) and Mediterranean (monitoring Soviet and Greek naval operations).

Weapon systems: AB 212ASW carries the Honeywell Mk 46 lightweight torpedo (see note below about anti-shipping missiles).

Personnel: 75 000 (total).

Remarks: Future plans include the upgrading of the AB 212ASW fleet with Ferranti Sea Spray radar for the British Aerospace Sea Skua missile for anti-ship as well as anti-submarine duties.

United Arab Emirates

Organisations: UAE Air Force.

Organisational structure: Formed from an amalgamation of emirate states along the Gulf coast which have pooled aircraft, maintenance and operational tasking. It is made up of Abu Dhabi, Ajman, Dubai, Fujaira, Ras al Khaima, Sharjah and Umm al Qaywayn, although both Abu Dhabi and Dubai retain their own air forces.

Command structure: Control of the UAEAF is thought to be under the Supreme Council of the UAE but other details are not known.

Air-capable ships: None.

Shore bases: Abu Dhabi, Ajman, Dubai, Fujeira, Ras al Khaia, Umm al Qaywayn.

Embarked aircraft: None.

Shore-based aircraft: Aerospatiale AS 332F Super Puma (4); Pilatus Britten-Norman Maritime Defender (2).

Units: No details available.

Recent operations: None reported.

Typical deployments: Supporting coastal surveillance and anti-smuggling patrols.

Weapon systems: AS 332F has provision for up to two AM 39 Exocet missiles for ASV role.

Personnel: 1500 (total).

Remarks: The full status of this force is not known.

United Kingdom

Organisations: Royal Navy (Fleet Air Arm); Royal Air Force: HM Coastguard; Department of Fisheries & Agriculture for Scotland (DAFS); Royal Marines.

Organisational structure: Fleet Air Arm is made up of 15 front-line and 7 second-line fixed-wing and helicopter squadrons, certain Type 42 destroyers have been earmarked to provide air defence cover for the UK Air Defence Region; the Royal Air Force is divided into commands, of which Strike Command operates the combat aircraft with maritime roles, with one subordinate group having the maritime responsibility, including the provision of air cover for warships within range; the air force also provides maritime combat rescue cover.

Command structure: Fleet Air Arm: the squadrons are subordinate to Flag Officer Naval Air Command (HQ Yeovilton) for manning, provision and training with some of the operational responsibility devolved to Flag Officer Third Flotilla, who also controls the operational strength of the Lynx squadrons. Royal Air Force: No 18 (Maritime) Group (HQ Northwood) provides long-range maritime patrol, anti-shipping strike, combat rescue and the co-ordination of maritime air tasks, under the command of the Air-Officer-Commanding the Group.

Air-capable ships: 3 *Invincible* class aircraft carriers (*Invincible*, *Illustrious*, *Ark Royal*); 12 Type 42 destroyers; 1 guided missile destroyer (*Bristol*); 7 Type 23 frigates (building); 4 Type 22 Batch 1 frigates; 6 Type 22 Batch 2 frigates; 4 Type 22 Batch 3 frigates;

British Aerospace Sea Harrier FRS1 is the standard Fleet fighter of the UK Royal Navy, armed with Aden 30 mm cannon (under fuselage) and up to four AIM-9L Sidewinder air-to-air missiles. The FRS2 improvement, with new radar and ASRAAM capability first flew in 1988 and will be in service by 1992. *(Paul Beaver)*

The first BAe Sea Harrier FRS2 made its maiden flight on 19 September 1988 and features the prominent Ferranti Blue Vixen radar.

6 Type 21 frigates; 2 Ikara *Leander* class frigates; 3 Exocet *Leander* class frigates; 4 *Leander* class towed-array conversions; 1 Gun *Leander* class frigates; 5 Seawolf *Leander* class frigates; 1 ice patrol ship (*Endurance*); 2 *Castle* class offshore patrol vessels (platforms only); 3 Ocean Survey craft; 2 seabed operations vessels; 3 *Ol* class large fleet tankers; 1 *Tide* class large fleet tanker; 5 *Rover* class small fleet tankers; 2 *Fort* class fleet replenishment ships; 2 *Fort* class auxiliary replenishment ships (building); 2 *Regent* class ammunition ships; 1 aviation training ship (*Argus*); 1 forward repair ship; 6 landing ships logistic.

The primary role of the Westland Lynx HAS2/3 small ship helicopter is anti-ship operations with the Sea Skua missile. In 1988 it was revealed that the Whittaker 'Yellow Veil' radar jammer was also fitted to the helicopter (see here outboard of a Sea Skua missile).

The primary medium anti-submarine warfare system is the Westland Sea King HAS5, armed with up to four Mk46 Mod 2 or Sting Ray torpedoes, GEC Avionics AQS-902 acoustics integration to process Jezebel sonobuoys; some units are also equipped with the Texas Instruments AQS-81 MAD. *(RN)*

Shore bases: Fleet Air Arm: Culdrose; Lee-on-Solent; Portland; Prestwick; Yeovilton; Fleetlands & Perth yards; RAF: Akrotiri (Cyprus); Finningley; Kinloss; Lossiemouth; Mount Pleasant/Kelly's Garden (Falklands); Valley.

Embarked aircraft: British Aerospace Sea Harrier FRS 1 (42) (9)*; Westland Sea King HAS 5 (75 + 7 on order); Westland Sea King HAS 6 (2); Westland Sea King AEW 2A (10); Westland Lynx HAS 2/3 (84) (7)*.

Shore-based aircraft: Fleet Air Arm: Aerospatiale Gazelle HT 2 (31); AMB-BA Falcon 20 (16); BAC Canberra TT 18 (9); BAe HS 125-700 (2 operated by RAF); BAe Sea Harrier T 4(N) (2); BAe Harrier T 4A (2); BAe Jetstream T 2 (16); BAe Jetstream T 3 (4); DH Chipmunk T 10 (14); DH Sea Devon

The first Sea King HAS6 under an evaluation flight for 824 Sqn Intensive Flying Trials Unit at Prestwick; the older Sea King HAS5 is pictured behind. *(RN)*

Fleet airborne early warning is provided by ten Westland Sea King AEW2 helicopters equipped with the Thorn EMI Electronics Searchwater radar. This helicopter is from 849A Flight embarked in HMS *Illustrious*. *(RN)*

Commando support operations are in the hands of the Sea King HC4 support helicopter which equips two front line Air Commando Squadrons; detachments include embarkation in the HM Ships *Fearless* (illustrated) and *Intrepid*. *(Paul Beaver)*

Daylight SAR and fleet support tasks are undertaken by Sea King HC4 helicopters from RNAS Portland on the English Channel coast. *(Patrick Allen)*

C 20 (3); DH Sea Heron C 1/C 4 (5); Hawker Hunter GA 11/PR 11 (11); Hawker Hunter T8C (9); Hawker Hunter T 7 (2); Hawker Hunter T8M (3); Westland Sea King HC 4 (33); Royal Air Force: Avro Shackleton AEW 2 (5); BAe Buccaneer S 2 (24); BAe Nimrod MR 2P (36); Westland Sea King HAR 3 (12); Westland Wessex HC 2/HU 5C (12); HM Coastguard: PBN Islander (1); Sikorsky S-61N (3); DAFS: Fokker F27 Mk 200 (1); Royal Marines: Aerospatiale Gazelle AH 1 (12); Westland Lynx AH 1 (12).

Units: RN: 702 (Lynx); 705 (Gazelle); 706 (Sea King); 707 (Sea King HC 4); 750 (Jetstream); 771 (Sea King); 772 (Wessex); 800 (Sea Harrier); 801 (Sea Harrier); 810 (Sea King); 814 (Sea King); 815 (Lynx); 819 (Sea King); 820 (Sea King); 824 (Sea King); 826 (Sea King); 829 (Lynx); 845 (Sea King HC 4); 846 (Sea King); 849 (Sea King AEW 2); RAF: No 8 Sqn (Shackleton); No 12 Sqn (Buccaneer); No 22 Sqn (Wessex HAR 2); No 42 Sqn (Nimrod); No 78 Sqn (Chinook/Sea King); No 84 Sqn (Wessex); No 120 Sqn (Nimrod); No 201 Sqn (Nimrod); No 202 Sqn (Sea King HAR 3); No 206 Sqn (Nimrod); No 208 Sqn (Buccaneer); SAR Trng Flt (Wessex).

Recent operations: Falklands; Gulf: escorts carry Lynx HAS 2/3 and auxiliaries have Wessex HU 5 embarked for vertrep and mine warfare duties.

Typical deployments: Fleet Air Arm: the two Invincible class aircraft carriers embark their respective Sea King HAS 5/Sea King AEW 2/Sea Harrier FRS 1 squadrons for deployment to the North Atlantic and Mediterranean, individual ships' flights for the 46 destroyers/frigates in commission are embarked as operations dictate, other helicopters are assigned to the ice patrol ship and ocean survey craft; auxiliaries embark helicopters for vertical replenishment and occasionally for operational front-line roles during exercise; shore-based aircraft deploy to overseas locations for training and trials; Royal Air Force: almost all the assets are home-based although they deploy as required; HM Coastguard/ DAFS: shore-based aircraft deployed around the coast as required.

Weapon systems: anti-shipping: British Aerospace Sea Eagle entered service with the Sea Harrier force

The UK Royal Navy has a requirement for up to 50 EH Industries Merlin helicopters, jointly manufactured by Agusta (Italy) and Westland (UK). The helicopter is expected to enter service in 1992. *(Westland)*

The UK Royal Air Force provides long-range maritime patrol aircraft for NATO and national defence duties. Nimrod MR2 aircraft are based at Kinloss (Scotland) and St Mawgan (Cornwall, England). *(RAF)*

in 1988 and gives the carrier-based aircraft an over-the-hòrizon role, Buccaneer S2 maritime strike aircraft will replace the TV-guided Martel with the Sea Eagle during 1988; Lynx helicopters are armed with up to four Sea Skua missiles, which were operationally tested prior to official in-service adoption in the Falklands, were tested again against range targets and hulks in August and September 1987 with 100% success, anti-submarine: RAF fixed-wing aircraft and naval helicopters can be armed with the Marconi Stingray/Honeywell Mk 46 Mod 2/Mk 44 Mod 0 lightweight torpedoes (nine torpedoes in the Nimrod weapons bay) or the British Aerospace Mk 11 Mod 3 depth bomb, selected aircraft are armed with the WE 177 nuclear free fall bomb; strike: Sea Harriers can be armed with SNEB 68 mm rocket pods/BL 755 cluster bombs/Mk 13/18 or Mk83 free-fall bombs; air defence four AIM-9L Sidewinder missiles carried by the Sea Harrier (provision for two Sidewinders by the Buccaneer S 2/Nimrod MR 2P) or four RN or LAU-69A pods for 19 70 mm or 51 mm rockets.

Long-range, day/night SAR for military and civilian operations around the UK coast is provided by the Westland Sea King HAR3.

Civilian aerial surveillance for fisheries protection duties is flown by Department of Agriculture & Fisheries for Scotland chartered Fokker F27 aircraft. Nordic Oil Services integrated the sensor package.

Personnel: Fleet Air Arm: 3400; Royal Air Force: 93 000 (total); HM Coastguard/DAFS: not available.

Remarks: Although the Royal Navy has three aircraft carriers in commission, there are only two front-line air groups; a third air group can be raised by depleting the training squadrons. In April 1987 it was announced that there is an intention to order 50 EH 101 (Merlin) helicopters for embarked use in the mid 1990s. In March 1988, the Wasp and Wessex HU 5 helicopters were withdrawn from service. A limited Sea Harrier update is underway and mainly concerns the radar and air-to-air missile capability; funds have not been made available for a full upgrade. A decision was made in early 1988 not to proceed with a full Lynx MK 8 update from 1989. Additional Sea King Mk 6 helicopters will be ordered in due course for delivery from 1990. Sea King HAS 6 Intensive Flying Trials Unit was formed at Prestwick in January 1988, supported by 819 and 824 Naval Air Squadrons based there; the Mk 6 has improved sonar, 360 degrees radar, better ASW processing and secure speech as well as data link.

Note: a Dormer 228 MPA has been delivered for fisheries surveillance and EE2 protection around the Falklands Islands.

US Naval Air Hierarchy

This list is correct to 1 May 1988 and was compiled by David Steigman of DMS Inc, a Jane's Information Group Company.

Central Staff & Command function:
Asst Chief of Naval Operations (Air Warfare): Vice Adm Robert Dunn. Commander, Naval Air Systems Command: Vice Adm Joseph Wilkinson Jnr. Director, Aviation Plans & Requirements: Rear Adm James Seely. Director of Operations, US Space Command: Rear Adm Jerry Breast. Chief of Naval Operation's Deputy Asst (Air Warfare): Adm Raymond Ilg. Commander, Fleet Air Mediterranean: Rear Adm Salvatore Gallo. Director, Space & Sensor Systems: Rear Adm Thomas Mattingly. Commander, Naval Air Test Center: Rear Adm John Calvert. Chief of Naval Air Training: Rear Adm David Morris. Program Director, Tactical Aircraft: Rear Adm George Strohsahl. Program Manager, V-22 Osprey: Brig Gen Harold Blot USMC. Command Director, NORAD Combat Operations Staff: Rear Adm Riley Mixson. Commander, Navy Space Systems: Rear Adm Donald Boecker. Director, Aviation Manpower and Training Division: Rear Adm Peter Cressy. Director, Naval Material Personnel Command: Rear Adm James Partington. Program Manager, P-3 Weapons Systems: Rear Adm William Vincent. Commander, Space & Naval Warfare Systems Command. Rear Adm Glenwood Clark Jnr. Vice Commander, Space & Naval Warfare Systems Command: Rear Adm John Weaver. Deputy Commander, Plans & Programs: Rear Adm Richard Friichtenicht. Asst Commander, Logistics & Fleet Support: Rear Adm John Kirkpatrick. Asst Commander, Space Technology: Rear Adm Thomas Betterton. Asst Commander, Systems & Engineering: Rear Adm Larry Blose. Director, Joint Cruise Missiles Project: Rear Adm William Bowes. Program Manager, LAMPS III/CV Helo: Rear Adm Robert Harrison.

US Atlantic Fleet:
Commander, Naval Air Force: Vice Adm Richard Dunleavy. Commander, Strike/Fighter Wing: Rear Adm Henri Chase III. Commander, Fighter Medium Attack/AEW Wings: Rear Adm James Taylor. Commander, Patrol Wings: Rear Adm John Yow. Commander, Helicopter Wings: Rear Adm Ronald Jesberg. Staff Commander, Naval Air Force: Rear Adm John Moriarity.

US Pacific Fleet:
Commander, Naval Air Force: Vice Adm John Fetterman Jnr. Commander, Light Attack Wing: Rear Adm Jeremy Taylor. Commander ASW Wing: Rear Adm John Adams. Commander, Patrol Wings: Rear Adm Phillip Smith. Commander, Medium Attack/EW Wing: Rear Adm Frederick Metz. Commander, Fleet Air Western Pacific: Rear Adm Bobby Lee. Commander, Fighter/AEW Wing: Rear Adm James Best. Staff Commander, Naval Air Force: Rear Adm Luther Schriefer.

Carrier Battle Groups:
Commander, Carrier Group One: Rear Admiral Kenneth Carlson. Commander, Carrier Group Two: Rear Admiral Richard Macke. Commander, Carrier Group Three: Rear Admiral David Rogers Commander, Carrier Group Four: Rear Admiral Jerome Johnson. Commander, Carrier Group Five: Rear Admiral Dennis Brooks. Commander, Carrier Group Six: Rear Admiral Leighton Smith Jnr. Commander, Carrier Group Seven: Rear Admiral Lyle Bull. Commnder, Carrier Group Eight: Rear Admiral William Dougherty Jnr.

United States of America

Organisations: US Navy: US Marine Corps; US Coast Guard.

Organisational structure: The US Navy is comprised of four active fleets: Second Fleet (Atlantic); Third Fleet (Eastern Pacific); Sixth Fleet (Mediterranean); Seventh Fleet (Western Pacific/Indian Ocean). Two of these fleets, designated the Atlantic and the Pacific Fleets, have their own naval air forces depending on the home basing ashore. These naval air forces are comprised of wings for patrol (maritime reconnaissance), tactical support, ASW (land-based and embarked), light attack, airborne early warning and electronic warfare. Each wing is made up of a number of squadrons which have the prefix V for fixed-wing and H for helicopter equipment. The suffix denotes role, for example HS = helicopter anti-submarine warfare and VA = fixed-wing attack. Squadrons are detached from the respective wings to form carrier air groups for the 15 fixed-wing aircraft carriers in commission and helicopters are detached to the 180 or so surface combatants which operate helicopters. The US Navy is the largest naval force in the world and the naval aviation element correspondingly supreme. The US Naval Reserve Air Forces are made up of six wings, two in support of aircraft carriers, two for maritime patrol, one with helicopters and the other for tactical support. Although strictly part of the US Navy, the US Marine Corps is considered here as a separate force although it is under the operational command of the US Navy, receiving funding from the US Department of the Navy's budget. As an arm of the USN, the USMC has its own naval air arm, based on a parallel structure to the navy and supporting amphibious assault operations. The US Coast Guard is subordinate to the US Department of Transportation in peacetime, but in tension/war, it becomes an arm of the USN. The USCG is divided into an Atlantic and Pacific

CANADA

CANADA

(Not to scale)

UNITED STATES OF AMERICA

Washington

MEXICO

Main USN AND USMC airfields

No.	Type	Airfield
1	NAS	Whidbey Island
2	NAS	Alameda
3	NAS	Moffett Field
4	NAF	Fallon
5	NAF	China Lake
6	NAS	Lemoore
7	NAS	Point Mugu
8	MCAS	Tustin
9	MCAS	El Toro
10	MCAF	Camp Pendleton
11	NAS	Miramar
12	NAS	North Island
13	NAF	El Centro
14	MCAS	Yuma
15	NAS	Kingsville
16	NAS	Corpus Christi
17	NAS	Chase Field
18	NAS	Dallas
19	NAS	New Orleans
20	NAS	Pensacola
21	NAS	Whiting Field
22	NAS	Meridian
23	NAS	Memphis
24	NAS	Atlanta
25	NAS	Cecil Field
26	NAS	Key West
27	MCAS	Beaufort
28	MCAS	New River
29	MCAS	Cherry Point
30	NAS	Oceana
31	NAS	Norfolk
32	MCAS	Quantico
33	NAF	Washington
34	NAS	Patuxent River
35	NADC	Warminster
36	NAS	Willow Grove
37	NAS	South Weymouth
38	NAS	Brunswick
39	NAS	Clenview
40	MCAS	Kaneobe Bay
41	NAS	Barbers Point

NADC = Naval Air Development Center
NAS = Naval Air Station
NAF = Naval Air Facility

MCAS = Marine Corps Air Station
MCAF = Marine Corps Air Facility

One of USS *Saratoga*'s combat air patrol Grumman F-14A Tomcat naval fighters on the port catapult. Note the undernose Northrop TV camera set, AIM-54 Phoenix, AIM-9 Sidewinder and AIM-7 Sparrow air-to-air missiles. With the two 1011 litres (267 US gal) fuel tanks, this is the standard high endurance CAP configuration. *(USN/PH1 W A Shayka)*

An F-14A Tomcat of Atlantic Fleet squadron VF-143 in typical close-range interception configuration of four AIM-7M, AIM-9M and the M61A1 20 mm six-barrel cannon. *(USN)*

organisation, each area commanded by a Rear Admiral USCG, reporting to a (4-star) admiral in Washington DC.

Command structure: USN: the basic unit is the squadron and when a number (of various roles) are embarked in an aircraft carrier they are called a Carrier Air Wing; the USMC can provide elements of a CAW when an assault landing is proposed, others are mainly anti-submarine warfare or sea control, a CAW is commanded by a Commander, helicopter wings/squadrons detach individual helicopters to sea in destroyers/cruisers/frigates; the USMC also operates with the squadron as the basic unit, subordinate to a Marine Air Wing when operational, subordinate to the respective Atlantic and Pacific Fleet Marine Forces; there are three MAWs supporting each active Marine Division: 1MAW at Kadena (Okinawa) with an air group at Kaneohe Bay (Hawaii), 2MAW at Cherry Point, 3MAW at El Toro; 4MAW is the reserve wing at Yuma; the order of battle is 30 tactical squadrons, 30 supported land force operations; seven for training and trials, plus a command support squadron of communications aircraft; squadrons are usually commanded by a Lieutenant Colonel; USCG: divided into 12 districts, each with one or more air stations, under the command of a Captain or Commander. Aircraft are assigned to stations rather than squadrons or flights.

Air-capable ships: US Navy: 3 *Iowa* class battleships (platform only); 3 *Nimitz* class nuclear-powered aircraft carriers (*Nimitz*, *Dwight D Eisenhower*, *Carl Vinson*); 1 Improved *Nimitz* class nuclear-powered aircraft carrier (*Theodore Roosevelt*) (2 building, 2 on order); 1 *Enterprise* class nuclear-powered aircraft carrier; 1 *John F Kennedy* class aircraft carrier; 3 *Kitty Hawk* class aircraft carriers (*Kitty Hawk*, *Constellation*, *America*); 4 *Forrestal* class aircraft carriers (*Forrestal*, *Saratoga*, *Ranger*, *Independence*); 2 *Midway* class aircraft carriers (*Midway*, *Coral Sea*);

This Atlantic Fleet squadron VPF-32 F-14A Tomcat shows the six AIM-54C missile configuration for maximum stand-off interception when the AWG-9 weapon control system can track 24 targets and attack six simultaneously at 170 mm (315 km).

Strike units in the USN/USMC are being equipped with the McDonnell Douglas F/A-18A Hornet, seen here immediately after launching from USS *Carl Vinson*. The aircraft is capable of air interception, anti-radiation target attack, anti-shipping attack, all-weather precision attack and reconnaissance tasks. *(USN/PHAN C R Solseth)*

4 *Virginia* class nuclear-powered cruisers (platform only); 2 *California* class nuclear-propelled cruisers (platform only); 1 *Truxton* class nuclear-propelled cruiser (platform only); 1 *Long Beach* class nuclear-propelled cruiser (platform only); 12 *Ticonderoga* class cruisers (12 building); 9 *Belknap* class cruisers; 4 *Arleigh Burke* class destroyers (building); 4 *Kidd* class destroyers; 10 *Coontz* class destroyers; 31 *Spruance* class destroyers; 5 *Forrest Sherman* class destroyers; 51 *Oliver Hazard Perry* class frigates; 46 *Knox* class frigates; 2 *Bronstein* class frigates; 2 *Blue Ridge* class command ships; 1 *Wasp* class helicopter carrier (4 building); 5 *Tarawa* class amphibious assault ships; 7 *Iwo Jima* class amphibious ships; 11 *Austin* class assault ships; 2 *Raleigh* class assault ships; 3 *Whidbey Island* class assault ships (5 building); 8 *Thomaston* class assault ships; 20 *Newport* class tank landing ships; 5 *Charleston* class amphibious cargo ships (platform only); 6 *Samuel Gompers* class tenders; 3 *Dixie* class tenders (platform only); 8 *Kilauea* class ammunition ships (platform only); 3 *Nitro* class ammunition ships (platform only); 2 *Suribachi* class ammunition ships (platform only); 7 *Mars* class combat stores ship, various auxiliary, special purpose,

McDonnell Douglas F/A-18A Hornet from squadron WFA-125 with general purpose bombs, demonstrating its strike role for the US Pacific Fleet.

Grumman A-6E Intruder, with conventional weapons, is the primary carrier-borne strike aircraft; A-6E update programmes will continue.

command and support ships including vessels of the Military Sealift Command and the Naval Reserve Force; USMC: operates from ships of the USN; USCG: 12 *Hamilton* class high endurance cutters; 13 *Bear* class medium endurance cutters; 16 *Reliance* class cutters (platform only); 1 *Storis* class cutter (platform only); 2 polar icebreakers; 2 *Polar Star* class icebreakers; 1 *Glacier* class icebreaker; 2 *Wind* class icebreakers; 1 *Mackinaw* class icebreaker (platform only).

Shore bases: USN: Atlantic Fleet: Beaufort, Brunswick, Cecil Field, El Centro, Fallon, Jacksonville, Key West, Norfolk, Patuxent River, Oceana, overseas: Guantanamo (Cuba), Rota (Spain), Sigonella (Italy). Pacific Fleet: China Lake, Lemoore, Miramar, Moffett Field, North Island, Point Mugu, Warminster, Whidbey Island, overseas: Agana (Guam), Atsugi (Japan) Cubi Point (Philippines), Kadena (Japan), Kamiseya (Japan), Misawa (Japan), USNR: Alameda, Andrews, Atlanta, Cecil Field, Dallas, Detroit, Glenview, Jacksonville, Ledmoore, Memphis, Miramar, New Orleans, Norfolk, North Island, Oceana, Patuxent River, Point Magu, South Weymouth, Whidbey Island, Willow Grove. Naval Air Training Command: Chase Field, Corpus Christi,

The Grumman KA-6D Intruder in-flight refuelling version from USS *Coral Sea* during Sixth Fleet operations in the Mediterranean. It is usual for each Intruder squadron to have four tankers on strength. Note the four 1136 litres (300 US gal) underwing tanks. *(USN/PH2 Rory Knepp)*

Grumman EA-6B Prowler specialist electronic warfare aircraft from the USN Atlantic Fleet.

Kingsville, Meridian, Pensacola, Whiting Field; USMC: Alameda, Cherry Point, El Toro, Kadena (Japan), Kaneohe Bay, New River, Quantico. USMCR: Alameda, Atlanta, Camp Pendleton, Dallas, Glenview, New Orleans, Norfolk, Santa Ana, South Weymouth, Whidbey Island, Willow Grove; USCG: Astoria, Barber's Point, Borinquen (Puerto Rico), Brooklyn, Cape Cod, Cape May, Chicago, Clearwater, Corpus Christi, Detroit, Elizabeth City, Houston, Kodiak, Los Angeles, Mobile, New Orleans, North Bend, Opa Locka, Port Angeles, San Diego, San Francisco, Savannah, Sitka, Tustin, Traverse City, Washington DC, Yuma.

Embarked aircraft: USN: Bell-Boeing HV-22A Osprey (50)*; 1 Boeing UH-46D/E/F Sea Knight (156)/(155)*; Grumman EC-2C Hawkeye (102) (18)*; Grumman A-6E Intruder (362); Grumman KA-6D Intruder (68); Grumman EA-6A Prowler (20); Grumman EA-6B Prowler (76); Grumman EA-6B Prowler III (9)*; Grumman F-14A/A/ plus Tomcat (476/32); Grumman F-14D Tomcat (127)*; Kaman SH-2F SeaSprite (167); Lockheed S-3A Viking (163); Lockheed S-3B Viking (22)*; McDonnell Douglas F/A-18A Hornet 302 (84)*; McDonnell Douglas F/A-18C/D (70); Sikorsky SH-3G/H Sea King (137); Sikorsky HH-60H Rescue Hawk (5) (13)*; Sikorsky

LTV A-7E Corsair II (seen here with AN/AAR-42 FLIR pods underwing) remains a potent carrier-borne attack aircraft but is now primarily assigned to USN Reserve units.

EC-124 (NKC-135A) on detachment to Norway. *(Patrick Allen)*

SH-60B Seahawk (183); Sikorsky SH-60F Oceanhawk (8) (168)*; Sikorsky RH-53D Stallion (12); Sikorsky MH-53E Sea Dragon (31) (7)*; Vought A-7E Corsair II (348).
USMC: Bell AH-1T SeaCobra (51); Bell AH-1W SuperCobra (44) (40)*; Bell UH-1N Twin Huey (204); Bell-Boeing MV-22A Osprey (552)*; Boeing CH-46D/E/F Sea Knight (269); Grumman A-6E Intruder (64); Grumman EA-6B Prowler (23); McDonnell Douglas F/A-18A Hornet (112); McDonnell Douglas F/A-18C Hornet (22); McDonnell Douglas RF-18D Hornet (83)*; McDonnell Douglas RF-4B Phantom (24); McDonnell Douglas A-4FM Skyhawk (84); McDonnell Douglas/British Aerospace AV-8B Harrier II (132) (42)*; Sikorsky CH-53D Sea Stallion (52); Sikorsky CH-53E Super Stallion (102).
USCG: Aerospatiale HH-65A Dolphin (90); Sikorsky HH-52A (20).

Shore-based aircraft: USN: Beech T-34C (318); Beech T-44A King Air (57); Beech UC-12B (82); Bell TH-57A/B/C Sea Ranger (165); Bell HH-1F Iroquois (20); Bell TH-1L (40); Bell UH-1M (12); Bell UH-1E/L/N Huey (54); Boeing NKC-135 (2); Boeing E-6A Hermes TACAMO (2) (3)*; Cessna T-47A Citation II (15); Convair C-131F Samaritan (20); Douglas KA-3B Skywarrior (19); General Dynamics F-16N (20)*; Gulfstream C-20D (2); Grumman C-1A Trader (20); Grumman C-2A Greyhound (38); Grumman TC-4C Academe (6); Grumman TE-2C Hawkeye (2); Grumman EC-2B Hawkeye (30); Israel Aircraft Industries F-21A Kfir (12); Lockheed C-130F/R Hercules (12); Lockheed EC-130Q TACAMO (16); Lockheed LC-130H Hercules (3); Lockheed RP-3A/D Orion (10); Lockheed EP-3E Orion (10); Lockheed P-3C Orion (256) (9)*; Lockheed US-3A (4); Lockheed KC-130 F/R/T (59); McDonnell Douglas C-9B Skytrain II (39); McDonnell Douglas F-4J/S Phantom II (48); McDonnell Douglas OA-4M Skyhawk (21); McDonnell Douglas C-9B Skytrain II (2); McDonnell Douglas TA-4M Skyhawk (22); McDonnell Douglas TA-4J Skyhawk (258); McDonnell Douglas F-4J/N/S Phantom II (200); McDonnell Douglas T-45A Goshawk (300)*; McDonnell Douglas F/A-18B/D (10); Northrop F-5E/F Tiger II (11); Northrop T-38A Talan (6); Rockwell CT-39E/G Sabreliner (12); Rockwell T-2C Buckeye (210); Rockwell T-39D Sabreliner (4); Sikorsky SH-3A Sea King (6); Sikorsky UH-60A (3); Sikorsky HH-60H (5); Sikorsky CH-53E Super Stallion (15); Vought TA-7C/EA-7L Corsair II (55).

Lockheed S-3A Viking carrier-borne anti-submarine warfare aircraft from USS *Eisenhower*. *(Lockheed)*

USMC: Beech UC-12B (17); Bell VH-1N Twin Huey (8); McDonnell Douglas/British Aerospace TAV-8B (28); Rockwell OV-10D Bronco (54); Sikorsky VH-3A (2); Sikorsky VH-53 (7); Sikorsky VH-60 (10)*.
USCG: AMD-BA HU-25A Gardian (41); Lockheed HC-130B/E/H (2/1/20) (5)*; Sikorksy HH-3F Pelican (31); Sikorsky HH-60J Rescue Hawk (35)*; Grumman VC-4A Gulfstream I (1); Grumman VC-11A Gulfstream II (1).

Units: USN: Atlantic Fleet: (Fleet Tactical Support Wing 1): VAQ-33 (special ops); VC-6 (drones); VC-8 (TA-4); VC-10 (TA-4): VQ-4 (EC-130Q); VRC-40

Grumman E-2C Hawkeye early warning and control aircraft which is embarked in USN strike aircraft carriers.

The internal layout of the E-2C Hawkeye showing the air control officer work station. *(USN)*

Small ship helicopter operations in older USN warships are flown by the Kaman SH-2F Seasprite helicopter. Standard equipment includes a Texas Instruments AQS-81(V)2 MAD and two 378 litres (100 US gal) fuel tanks.

Evaluation prototype YSH-3G Supersprite, powered by two General Electric T700 engines to give the airframe an operational role fo, the late 1990s.

(C-1/CT-39); VRF-31 (ferry unit); VXN-8 (P-3); (Patrol Wing 5); VP-8; VP-10; VP-11; VP-23; VP-26; VP-44; VP-58; (P-3B/C); (Patrol Wing II): VP-5; VP-16; VP-24; VP-45; VP-49; VP-58; (P-3B/C): VP-30; (P-3B/C): (ASW Wing Atlantic); VS-22; VS-24; VS-28; VS-30; VS-31; VS-32; (S-3A): (Helicopter ASW Wing I): HS-1; HS-3; HS-5; HS-7; HS-9; HS-11; HS-15; HS-17 (Sea King); (Helicopter Tactical Wing I); HC-4; HC-6; HC-8; HC-16; (CH-46D/E): HM-12; HM-14; HM-15; HM-16 (RH-53D/CH-53E/MH-53E); (Helicopter Sea Control Wing 1); HSL-30; HSL-32; HSL-34; HSL-36; HSL-42; (SH-2); VX-1 (development unit); (Tactical Wings Atlantic): (AEW Wing 12); VAW-120; VAW-121; VAW-122; VAW-123; VAW-124; VAW-125; VAW-126; VAW-127; VAW-132 (E-2); (Fighter Wing 1); VF-11; VF-14; VF-31; VF-32; VF-33; VF-41 (F-14A/C); VF-43 (Kfir/TA-4); VF-74; VF-84; VF-101; VF-102; VF-103; VF-142; VF-143 (F-14A/C); (Light Attack Wing 1); VA-12; VA-15; VA-37; VA-45; VA-46; VA-66; VA-72; VA-81; VA-82; VA-83; VA-86; VA-87; VA-105; (A-7E); VFA-106; VFA-131; VFA-132; VFA-136; VFA-137 (F/A-18); VA-174 (A-7E); (Medium Attack Wing 1); VA-34; VA-35; VA-42; VA-55; VA-65; VA-75; VA-85; VA-176 (KA-6D/A-6E); VQ-2 (EP-3); VR-24 (C-130; CH-53E). Pacific Fleet: HC-1 (Sea King); HC-3; HC-5; HC-11 (CH-46D/E): VAQ-34 (EW): VC-1; VC-5 (various); VQ-1 (EP-3); VQ-3 (EC-130Q); VRC-30 (C-12/C-2/CT-39); VRC-50 (C-2/C-130): VX-4; VX-5 (various); VXE-6 (LC-130/UH-1N): (ASW Wing Pacific): HS-2; HS-4; HS-6; HS-8; HS-10; HS-12; HS-14 (Sea King); HSL-31; HSL-33; HSL-35; HSL-37 (SH-2); HSL-41; HSL-43 (SH-60B); VS-21; VS-29; VS-33; VS-35; VS-37; VS-38; VS-41 (S-3A); (Fighter/AEW Wing Pacific): VAW-110; VAW-112; VAW-113; VAW-114; VAW-115; VAW-116; VAW-117 (E-2B/C): VF-2; VF-21; VF-24; VF-51; VF-111; VF-114; VF-124 (F-14A/C); VF-126 (A-4); VF-151 (F-4); VF-154 (F-14); VF-161 (F-4); VF-211; VF-213 (F-14A/C); Fighter Weapons School (Top Gun) (A-4/TA-4/F-5E); (Light Attack Wing Pacific): VA-22; VA-56; VA-93; VA-94; VA-97 (A-

A Pacific Fleet squadron HS-4 Sikorsky SH-3H Sea King from USS *Carl Vinson* recovers its Bendix ASQ-13E dipping sonar during routine training. *(USN)*

7E); VFA-25; VFA-113; VFA-125; VFA-195 (F/A-18); VA-122; VA-146; VA-147; VA-192 (A-7E); VA-127 (A-4/TA-4); (Medium Attack/EW Wing Pacific): VA-52; VA-95; VA-115; VA-128; VA-145; VA-165; VA-196 (Intruder); VAQ-129; VAQ-130; VAQ-131; VAQ-132; VAQ-133; VAQ-134; VAQ-135; VAQ-136; VAQ-137; VAQ-138; VAQ-140 (Prowler); (Patrol Wing 1): controls US-based aircraft deployed to Far East; (Patrol Wing 2); VP-1; VP-4; VP-6; VP-17; VP-22 (P-3B/C); (Patrol Wing 10); VP-9; VP-19; VP-31; VP-40; VP-46; VP-47; VP-48; VP-50 (P-3B/C); USN support: Commander Logistics Support Wing (Gulfstream C-20D); USMC: (1MAW); VMFA-212; VFMA-232; VFMA-235; VFMA-312; VMA-211;

The US Navy's next generation lightweight air-launched torpedo is the Honeywell Mk50 Barracuda, seen during trials from an SH-3D Sea King.

Modern USN cruisers, destroyers and frigates will embark the Sikorsky SH-60B Seahawk helicopter as the Light Airborne Multi-Purpose System's air vehicle. It is equipped with IBM APS-124 search radar and provision for two torpedoes and/or the Penguin Mk3 anti-ship missile.

USN strike aircraft carriers will be equipped with the Sikorsky SH-60F Oceanhawk and the Bendix ASQ-13F dipping sonar to replace the SH-3H in due course. The primary task of this helicopter is to protect the inner zone around the aircraft carrier.

Sikorsky RH-53D Sea Stallion aerial mines and countermeasures helicopter towing an MCM sled. *(USN)*

VMA-332; VMGR-152; HMH-463; HMH-361; HML-367; HMM-163; HMM-165; HMM-262; HMH-265; (2MAW); VMFA-115; VMFA-122; VMFA-251; VMFA-451; VMA-223; VMA-224; VMA-231; VMA-331; VMA-533; VMA-542; VMO-1; VMGR-252; HMA-269; HMH-362; HMH-461; HML-167; HMM-162; HMM-261; HMM-263; HMM-264; (3MAW); VMFA-314; VMFA-323; VMFA-531; VMA-121; VMA-214; VMA-242; VMA-513; VMO-2; VMGR-352; HMA-169; HMA-369; HML-267; HMM-161; HMM-164; HMM-268; HMM-365; USMCR: (4MAW): VMFA-112; VMFA-321; VMA-124; VMA-131; VMA-133; VMA-134; VMA-142; VMA-322; VMO-4; VMGR-234; HMA-773; HMH-769; HMH-772; HMH-777; HML-770; HMM-764; HMM-767; HMM-774.

The Sikorsky MH-53E Sea Dragon will replace the RH-53D in the near future; it has an enhanced capability including long endurance and towing power with its three General Electric T64 engines.

Boeing UH-46D Sea Knight tandem-rotor combat support and utility helicopter aboard HMS *Illustrious*, the UK Royal Navy aircraft carrier during a cross-deck exercise. *(RN)*

Recent operations: 1987: aircraft from *Constellation* patrolled the Gulf convoy routes and Sikorsky RH-53D Stallion aerial mines countermeasures aircraft were embarked in various ships, including the command ship *La Salle*, to clear mines allegedly laid by Iran.

Typical deployments: USN: as a rule 12 fixed-wing aircraft carriers are in commission, the bulk with the Second and Third Fleets, one to three each with the Sixth and Seventh Fleets; in addition there are 12 amphibious warfare carriers (usually embarking USMC aircraft), about 30 cruisers, 70 destroyers and 110 frigates, most of which have a single detached helicopter flight embarked; a typical USN carrier air wing in a nuclear-powered aircraft carrier (CVN) would consist of two fighter squadrons (VF), three attack squadrons (VA), and electronic warfare unit (VAQ), an airborne early warning squadron (VAW), a fixed-wing ASW squadron (VS) and a helicopter ASW unit (HS); USNR: supports the USN as required; USMC: overseas deployment by sea in a task group would involve a Marine Expeditionary Brigade with a mixed air group of 110 fixed-wing aircraft and 120 helicopters, a Marine Amphibious Unit is supported by 6 fixed-wing aircraft and 30 helicopters; USMCR: supports the USMC as required; USCG: helicopters operate with the cutters on patrol from the Arctic to the Antarctic, long-range maritime patrol aircraft carry out surveillance tasks, especially law enforcement duties to prevent drug smuggling.

Weapon systems: USN/USMC: anti-shipping/strike

The first Sikorsky HH-60H Helicopter Combat Support aircraft for USN combat rescue and special operations transport.

Lockheed P-3C Orion Update III prior to delivery. Note the chin-mounted AAS-36 infra-red detection set for night/bad weather identification.

weapons include the provision of Harpoon for the P-3C Orion Update II/III/IV, a weapon also carried by the two squadrons of the B-52 D/G bombers allocated by the USAF, the carrier-borne Hornet can carry 7,700 kg of iron bombs, the Corsair II is armed with the Walleye II missile and/or Mk57 nuclear weapons; the older Orions still have provision for the 127 mm rocket; the Penguin missile for the SH-60B Seahawk now appears doubtful due to funding restrictions; for anti-submarine warfare, the Orion can carry eight Mk 46 torpedoes in the weapons bay and four externally, or various combination loads, including the 454 kg mine and Mk 101 nuclear depth charge, the Sea King and SeaSprite can be armed with the Mk 46 or MK44 lightweight torpedo and are expected to be armed with the Mk 50 ALWT when it enters service; air defence weapons include the AIM-9 Sidewinder (in several models, including

The prototype Boeing E-6A TACAMO communiations aircraft prior to delivery to the USN. When operational in the 1990s, the E-6 will be the US Navy's largest aircraft. *(Boeing)*

Harpoon anti-ship missile launch from a P-3C Orion, part of the aircraft's secondary role of anti-surface vessel warfare.

a helicopter system for the SuperCobra), the AIM-7 Sparrow and the Phoenix for long-range air superiority, fitted to the Tomcat; cannon are also fitted to a number of types, including the Corsair II and the AV-8B Harrier II; future programmes include Hughes Aircraft AMRAAM which by August 1987 had completed 18 successful test firings from a total of 21 using the F/A-18 fighter with look-down, shoot-down capability as a test bed. USCG: all aircraft are unarmed but carry various surveillance equipment.

Personnel: USN/USMC: 61 700; USCG: not available.

Remarks: The continuation of the nuclear-powered aircraft carrier and the new amphibious helicopter carrier programmes will ensure that the US Navy and US Marine Corps maintain their naval aviation assets at sea and shore-based into the next century. In October 1988 President Reagan approved a multi-year purchase of 63 AV-8Bs and nine TAV-8Bs for the USMC for FY 1989–91 funding. Primary training

Boeing Aerospace is modifying a number of P-3C Orion to Update IV standard to address the problems of newer and quieter nuclear-powered submarines. The first prototype system is due to be delivered in mid 1990 and some 205 systems will be procured after full-scale engineering development in 1992.

The US Navy announced in September 1988 that it will go ahead with the US-3A programme to replace the C-2A Greyhound carrier on-board delivery (COD) aircraft.

will continue to be carried out by the Beech T-34C, a further 19 will be delivered in 1989-90 and the first of 300 McDonnell Douglas T-45A (BAe Hawk) was flown in December 1987. The USCG has a continued aviation role which will be devoted more to law enforcement than safety at sea.

Future programmes

Advanced Tactical Aircraft The USN intends to have 450 advanced aircraft available for service in the late 1990s. Northrop (prime) and McDonnell Douglas (main sub) have been contracted to produce design studies for ATA.

Selected Navy Aviation Programmes

Advanced Capability Prowler: The EA-6B Prowler aircraft is being given increased capability after experimental work by Grumman, the US Naval Air Systems Command and NASA under the Vehicle Improvement Programme which will include the installation of the Pratt & Whitney J52-P-409 engine for increased single-engine performance from carrier decks.

Airship Programme: Although the USN Airship Programme was eliminated from the US budget in February 1988, the project could well be allocated to the Defense Advanced Research Projects Agency if support is received from the US Customs and US Coast Guard. The programme is a joint programme between Westinghouse and Airship Industries.

Combat Rescue: Nine Sikorsky HH-60H Rescue Hawks will be acquired for USN combat (search &) rescue operations and special warfare. A full order of 18 aircraft is expected by 1990.

LRAACA: The Long-Range Air ASW Capability Aircraft programme is currently under review by the USN and the Lockheed suggestion of an enlarged and enhanced version of the Orion was selected in October 1988. Other proposals were for the development of the Boeing 757 and a prop-fan engined McDonnell Douglas MD-89 airliner project. The avionics fit would be to Update IV standard as contracted to Boeing Aerospace for the P-3C programme. The US Navy sought competition bids for 125 P-3G Orions for 1990-94 in 1988.

Night Attack Navigation System update: Grumman Corporation has been awarded a major contract to improve the A-6E Intruder's night attack capability. An essential part of the NANS is a forward looking infra-red (FLIR) system developed by GEC Avionics for night and poor weather operations. This will significantly increase the aircraft's utilisation, reduce pilot workload and enhance the single pass attack capability of the aircraft. The FLIR sensor uses the Thermal Imaging Common Module Class II and the display is compatible with night vision goggles also being supplied by GEC Avionics.

Shore-based Grumman C-2A Greyhound aircraft are used to support USN carrier battle groups with on-board deliveries. *(USN/PHC Dunivan)*

USN carrier groups occasionally take the EA-3B Skywarrior to sea for special electronic warfare operations such as the US raid on Libya.

Oceanhawk CV-HELO: The Sea King replacement programme began operational flying trials in June 1987 and the USN has a requirement for 175 aircraft. Sikorsky Aircraft is the prime systems integrator. The helicopter will have an average operational endurance of four hours and be capable of carrying low frequency dipping sonar and two Honeywell Mk 50 Barracuda lightweight torpedoes.

Osprey: The US Department of the Navy is the programme lead for the Bell-Boeing Tilt-Rotor programme which is designed to enhance the US Marine Corps' future assault transport operations. The USMC needs 552 MV-22A, the USN wants 50 HV-22A for combat rescue and perhaps 300 SV-22A as a replacement for the S-3A/B Viking anti-submarine warfare aircraft. The first USMC units will be operational in 1991 and the remainder of the Ospreys will be delivered by 2001. Supplemented CH-46D/E Sea Knight helicopters will be transferred to the USN for replenishment and other support duties.

P-3 Update IV: This is the anti-submarine warfare capability update programme for the Orion long-

Advanced carrier-borne aircrew training will be carried out in the McDonnell Douglas/British Aerospace T-45 Goshawk from 1990.

range maritime patrol aircraft. The contract has been awarded to Boeing Aerospace and involves acoustic and non-acoustic sensors, improved communications equipment, distributed processing, data bus architecture and programmable full-colour displays. Update IV received Navy Systems Acquisition Review Council approval on 7 November 1984, in November 1985 the contractor received Demonstration & Validation authority and system demonstrations were begun in October 1986. On 10 July 1987 Boeing Aerospace was awarded a $244 million full-scale engineering development contract by the US Naval Air Systems Command for a prototype aircraft in 1990. The contract includes four firm-fixed-price options of $830.1 million for 140 Update IV systems beginning in 1991. The USN plans to buy 205 systems over a six-year period with 80 sets being retro-fitted into P-3C Orions to replace Update II equipment and the bulk being set aside for the LRAACA programme.

Seahawk Update: The SH-60B Seahawk LAMPS III air vehicle will be re-engined with the General Electric T700-GE-401C to give greater endurance, better safety margins and to allow for the carriage of the Grumman/Kongsberg Penguin Mk 2 anti-shipping missile. An additional station has been designed to give the option of a missile and two tanks. Other improvements include the installation of Global Positioning System AN/URC-182 V-/VHF radios and provision for the carriage of the Honeywell Mk 50 Barracuda lightweight torpedo. Most of the planned 204 helicopter fleet will be upgraded with the new equipment from late 1988 in a rolling, block modification programme.

Signals Intelligence Aircraft: The USN has awarded Lockheed Aeronautical Systems a $66 million contract to develop the ES-3A prototype carrier-borne SIGINT aircraft and a $88 million option for 15 additional aircraft. There is a requirement for 24 aircraft by 1990/91. The US Naval Avionics Centre will act as electronics integrator and the new aircraft will be deployed to newly-formed units VQ-5 and VQ-6 at Agena and Rota.

T-45 Goshawk: The T-45TS (training system) programme requires 300 aircraft, 32 flight simulators, 49 computer-aided instructional systems, four training integration systems mainframe computer, 200 terminals and academic materials sufficient to allow the training of up to 600 USN pilots per year. T-45TS replaces the existing training system and two separate training aircraft, the T-2C Buckeye and TA-4J Skyhawk; according to British Aerospace and McDonnell Douglas, the much lower running costs of the T-45A Goshawk training system and the increased use of simulation and advanced academic instruction will result in a cost reduction per student of nearly 50 per cent during this phase of their training. The Goshawk will enter service at Kingsville NAS, Texas, in late 1990 and be

Beechcraft T-34C Mentors are used for primary pilot training and photographed at a land-away at a local airfield in Georgia. *(Paul Beaver)*

The Northrop F-5E Tiger II is flown for aggressor training. (USN)

deployed to Chase Field, Texas and Meridian, Mississippi when aircraft becomes available. The contract for the first 12 airframes was valued at $429.4 million (December 1987) and FY 1989 budget includes a further $517 million of 24 additional aircraft and associated training systems; production of the Goshawk will peak in 1993 at 48 aircraft per year and continue at this rate until 1997.

TACAMO: Boeing Aerospace is developing a survivable airborne communications system for the US Navy's submarine-launched ballistic missile (SLBM) force based around the E-6A Hermes aircraft. It will serve as a back-up link between the SLBM force and the national command authorities. In the event of the ground-based system being destroyed, the command authority would pass to the E-6A under the aegis of the Take Charge and Move Out (TACAMO) programme. The E-6A is based on the E-3A Sentry AWACS (Airborne Warning and Control System) aircraft and will carry the USN's airborne very low frequency (AVLF) communications system in a hardened Boeing 707-320B commercial aircraft,

From 1985, some USN aggressor squadrons were equipped with the Israeli Aircraft F-21A Kfir. *(USN)*

powered by four General Electric/SNECMA CFM-56 engines. Long-lead funds for the first aircraft were provided in the FY 1984 budget and FY 1986 appropriated funds for the production of two aircraft and long-lead items for three others as part of a total contract award of $1,500 million. The USN has a requirement for 15 aircraft. The prototype E-6A first flew in February 1987 and deliveries begin in early 1989; the last aircraft is expected in late 1990.

Tomcat Update: Grumman is working on the production of the F-14 (Plus) and the F-14D versions of the Tomcat swing-wing air superiority fighter. In April 1988, the company was awarded a $20.6 million increment under the contract for full-scale development which is expected to be completed in the spring of 1990. A simulator programme is also underway at Grumman, which is expected to be finished in June 1996.

Undergraduate Jet Pilot Training System: The T-45A Goshawk jet trainer is the aircraft element of the T-45 Training System which itself forms the core of the UJPTS. McDonnell Douglas is the prime contractor and systems integrator with British Aerospace acting as the major sub-contractor with Rolls-Royce who manufacture the Adour Mk 861-40/F405-RR-400 engines for the programme. Simulation technology is provided by Honeywell.

Various training and experimental USN units fly the TF-18A Hornet, seen here during trials with the Texas Instruments HARM anti-radiation missile which also arms the USN A-7E Corsair II.

McDonnell Douglas A-4M Skyhawk is now a training and utility jet.

Primary helicopter training is carried out by a small number of Iroquois-derived variants, including the TH-1L.

The bulk of helicopter pilot training is now carried out by the Bell TH-57A/B/C SeaRanger, mainly at Pensacola NAS, Florida.

Local SAR and other emergency tasks are carried out by the UH-1N TwinHuey. *(Bell)*

USS *Guadalcanal*, *Iwo Jima* class, with a composite USMC helicopter squadron but also capable of carrying the AV-8B Harrier II for limited operations.

USMC light attack units now fly the McDonnell Douglas/BAe AV-8B Harrier II for 'mud moving operations'.

The Bell AH-1W SuperCobra is armed with the Rockwell Hellfire missile for the anti-armour role.

Most Bell AH-1T SeaCobras will be converted to AH-1W standard in due course.

A limited number of Rockwell OV-10D Bronco multi-purpose aircraft are still flown by the USMC even though the type first entered service more than 20 years ago. It is capable of carrier operations from amphibious warfare ships. *(Paul Beaver)*

Heavy lift and assault operations are flown by the Sikorsky CH-53E Super Stallion, seen here carrying an 11.25 tonnes trestle bridge centre section.

All future amphibious assault operations are due to be flown by the Bell-Boeing V-22 Osprey tilt-rotor seen here at its roll-out in May 1988. *(Paul Beaver)*

Most ships operating with USMC amphibious units embark the Bell UH-1N.

Lockheed C-130H Hercules from the US Coast Guard detachment at Clearwater, Florida used for long-range patrol and rescue operations in the Caribbean Sea.

Sikorsky will supply the HH-60J Rescue Hawk, a medium range recovery helicopter to the US Coast Guard to replace the HH-3F Pelican from 1990.

A small number of Sikorsky HH-52A Seaguard helicopters remain in service and will be replaced by the HH-65A by 1990. *(Paul Beaver)*

Short range recovery tasks for the US Coast Guard are performed by the Aerospatiale HH-65A Dolphin, based on the SA 366G1 version of the Dauphin helicopter.

Sikorsky HH-3F Pelican being washed down at Clearwater CGAS after an overwater recovery.

USSR

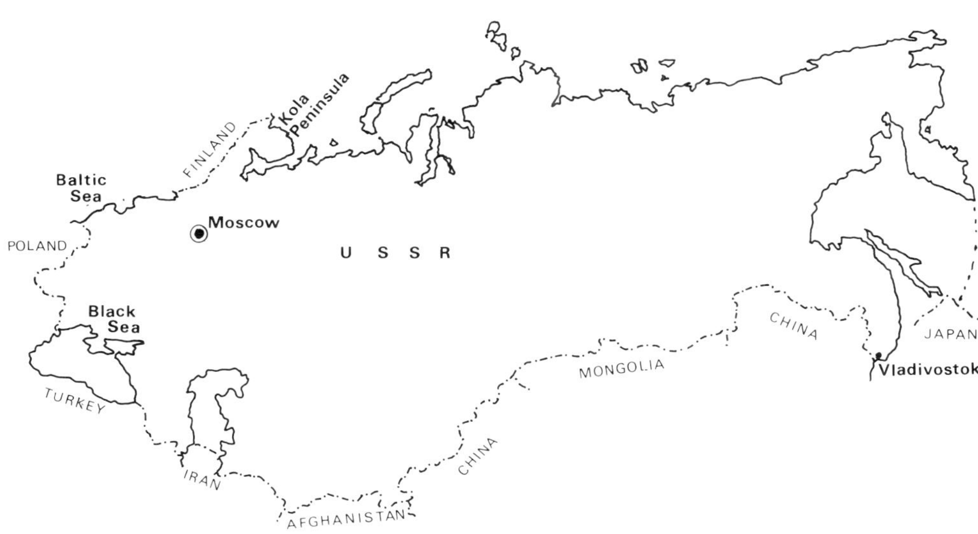

Organisations: Soviet Naval Aviation (AV-MF).

Organisational structure: No precise details are known, although it is understood that the navy controls all maritime operations; it is estimated that there are about 70 000 personnel, 1000 combat aircraft and 300 transport and second-line types. The four Soviet Red Banner Fleets (Northern, Baltic, Black Sea and Pacific) each have a naval air force attached to them. The four *Minsk* class aircraft carriers have 12/15 VTOSL Yak-38 fighters and 20/25 Ka-25 helicopters embarked and the two helicopter carriers operate about 20 helicopters (Ka-25, Ka-27 and occasionally Mi-14 'Haze-B')/ The naval infantry is supported by AV-MF helicopters. Initial training

In June 1988 the fourth *Kiev* class aircraft carrier, *Baku*, made its appearance in the Mediterranean. Unlike earlier ships in the class, it has a redesigned superstructure for a phased array radar.

Novoroosiysk operating in the Pacific Ocean during 1986. Note the two-seat Yak-36 'Forger B' embarked for unknown trials. *(USN)*

The aircraft carrier, *Kiev*, showing the large selection of missiles aboard and Kamov Ka-27 'Helix' helicopters aboard. At first glance, it appears that the single helicopter on the angle-deck is the duty stand-by helicopter but the presence of a higher alert state Ka-27 forward is confusing.

is carried out by the semi-military DOSAAF organisation and the Frontal Aviation (air force).

Command structure: No details are known other than the basic unit is the regiment.

Air-capable ships: 4 aircraft carriers (*Kiev*, *Minsk*, *Novorossyisk*, *Baku*); 1 nuclear-powered aircraft carrier (building); 2 helicopter carriers (*Moskva*, *Leningrad*); 3 nuclear-powered cruisers (*Kirov*, *Frunze*, *Kalinin*); 3 *Slava* class cruisers; 7 'Kara' class cruisers; 10 'Kresta-II' class cruisers; 4 'Kresta-I' class cruisers; 2 'Modified Sverdlov' class cruisers; 8 *Udaloy* class destroyers (some building); 8 *Sovremennyy* class destroyers (6 building); 8 'Kanin' class destroyers; 3 'Modified Kanin' class destroyers; 2 'Krivak-III' class frigates; 4 *Arktika* class ice breakers (1 building); 6 large icebreakers; 4 *Kapitan Sorokin* class icebreakers; 8 *Ivan Susanin* class armed icebreakers; 4 *Sibir* class missile range ships; 1 *Marshal Nedelin* class missile range ship; 3 *Alesha* class minelayers (platform only); 2 *Ivan Rogov* class landing ships (1 building); 6 *Don* class command ships; 7 'Ugra' class depot ships; 2 *Malina* class repair ships; 1 submarine rescue ship; 2 *Ob* class hospital ships; 1 missile support ship.

Shore bases: Details not available (see map).

Embarked aircraft: Kamov Ka-25 'Hormone A' (100); Kamov Ka-25 'Hormone B' (20); Kamov Ka-25 'Hormone C' (10); Kamov Ka-27 'Helix A' (50); Kamov Ka-27 'Helix B' (20); Mil Mi-14 'Haze B' (20); Yakolev Yak-38 'Forger A' (56).

Kiev in the Mediterranean Sea with two of her deck park Yak-36s cocooned against the weather. Note the UK-style all-wheel drive tractor under the superstructure.

The first Soviet jet shipborne fighter was the Yak-36 'Forger A' which operates in the vertical take-off short landing mode. In *Kiev* class aircraft carriers, it is common for two of the aircraft to be deck parked.

A vertical view of *Moskva* shows the helicopter carrier's four landing spots. Kamov Ka-25 'Hormone A' helicopters are ranged on deck.

Shore-based aircraft: Antonov An-12 'Cub various' (40); Beriev Be-12 'Mail' (80); Ilyushin II-20 'Coot A' (10); Ilyushin II-38 'May' (50); Mil Mi-8 'Hip' (50); Mil Mi-14 'Haze A' (100); Sukhoi Su-17M 'Fitter C/D' (74); Tupolev Tu-16 'Badger' (400); Tupolev Tu-22 'Blinder' (25); Tupolev Tu-22M/26 'Backfire B' (105); Tupolev Tu-95/Tu-142 'Bear series' (95); Transports: Antonov An-2, An-12, An-24, An-26; Ilyushin II-14; Mil Mi-4 'Hound'; Mil Mi-8 'Hip C'; Trainers: Aero L-29 Delfin; Aero L-39 Albatross; Sukhoi Su-17U; Tupolev Tu-22U; Yakolev Yak-18; Yakolev Yak-38UV.

Units: No accurate data available.

Recent operations: None reported.

Typical deployments: Most Soviet principal warships

Most Ka-25 'Hormone A' anti-submarine warfare helicopters are embarked in cruisers, destroyers or large frigates. The chin radome houses the 'Big Bulge' I/J-band radar. *(USN)*

For newer ASW escorts and the first fixed-wing aircraft carriers, the Kamov bureau developed the Ka-27 'Helix', seen here in its A or standard ASW version. *(USN)*

now embark helicopters and the aircraft carriers have the 'Forger' operational so may be encountered on any ocean; long-range maritime patrol aircraft are known to be based in Angola, Cuba, Ethiopia, Guinea-Bissau, Mozambique, Vietnam as well as with the four main fleet organisations. It is possible for the naval air fleets to operate over most oceans in the world and the use of in-flight refuelling extends the land-based aircraft range across the Atlantic, Indian and Pacific Oceans.

Weapon systems: Anti-shipping: fixed-wing aircraft carry a range of stand-off and cruise anti-ship missiles

In 1987, it was revealed that there was a naval assault version of 'Helix', the C version.

Long-range overwater reconnaissance operations are flown predominantly by the Tu-142 'Bear D' of which about 40 are thought to be in service with the Northern Fleet.

Two versions of the intermediate range Tu-16 include the 'Badger A' tanker variant which is shown in-flight refuelling the 'Badger C' anti-ship version. The latter is capable of carrying a single AS-2 'Kipper' missile and is in service with the Baltic, Black Sea, Northern and Pacific Fleets.

Modified for long-range electronic warfare/reconnaissance operations, the Tu-16 'Badger E' also carries photographic equipment.

The Tu-16 'Badger G' has been modified to carry one (as here) or two AS-6 'Kingfisher' missiles and is in service with the Black Sea, Northern and Pacific Fleets.

ranging from the AS-7 'Kerry' or AS-9 weapons carried by the Su-17M to the two AS-6 'Kingfisher' missiles of the Tu-16 'Badger G', the single AS-4 'Kitchen' of the Tu-22, and the single AS-4 or two AS-6 of the Tu-26. The Tu-26 can also carry more than 16 000 kg of free-fall general purpose, incendiary or nuclear bombs; the 'Bear' series have been credited with 20 000 kg of bombs, including three 50-megaton thermo-nuclear weapons and a major threat is posed to NATO shipping by the 'Blackjack' armed with four 50-megaton nuclear weapons or up to eight AS-15 anti-shipping, air-launched cruise missiles; carrier-borne Yak-38 could carry two AS-7 missiles or about 4 000 kg of free-fall bombs and/or 55 mm rocket pods; anti-submarine: the helicopter assets are armed with a variety of lightweight torpedoes, depth bombs (including nuclear types) and other charges although details are not known; air defence: carrier-borne Yak-38 aircraft can carry a combination of the AA-2 'Atoll', AA-2-2 'Advanced Atoll' or AA-8 'Aphid' as well as two 23 mm cannon pods, air assault: Mi-8 helicopters can be used for assault operations, armed with four AT-2 'Swatter' wire-guided missiles or six 32-round 55mm rocket pods or six FAB-250 250kg free-fall bombs.

Personnel: 70 000.

Remarks: The Soviet naval air force is growing steadily as the number of Soviet warships with flight decks and hangers increases, and the policy of world-wide naval activity continues. Because of its lack of overseas bases the Soviet Union uses a number of friendly country facilities to spread its long-range, shore-based maritime patrols to cover the South Atlantic, Indian and southern Pacific Oceans. New aircraft types continue to be introduced, including the Kamov Ka-27 helicopter which will replace the ASW version of the Ka-25 by 1992 and the Mi-14 in its mine countermeasures version. To assist in the role of the latest Soviet aircraft described above, the following chart has been prepared.

Type	Suffix	Role
'Backfire'	A	strike
'Backfire'	B	strike
'Backfire'	C	strike
'Badger'	C	strike
'Badger'	D	recce
'Badger'	G	strike
'Badger'	H	ECM
'Bear'	D	Recce
'Bear'	F	ASW
'Bear'	F	Mod II ASW
'Bear'	F	Mod III ASW
'Blinder'	C	PR/Recce
'Haze'	A	ASW
'Haze'	B	AMCM
'Haze'	C	SAR/utility
'Helix'	A	ASW
'Helix'	B	assault
'Helix'	C	civilian
'Helix'	D	SAR/utility
'Hormone'	A	ASW
'Hormone'	B	OTHT
'Hormone'	C	SAR/utility

Tupolev Tu-26/Tu-22M 'Backfire B' is a swing-wing bomber which the Soviets claim is not for long-range operations; during the SALT 2 talks the overnose-mounted in-flight refuelling probe was removed.

Beriev Be-12/M-12 'Mail' medium range ASW amphibian are operational with Black Sea and Northern Fleets; occasionally they are also observed in the Baltic Sea. *(BMVg Bonn)*

Ilyushin Il-38 'May' is a combined long-range maritime patrol and anti-submarine warfare aircraft. The undernose radome carries an unidentified search radar.

Ilyushin Il-20 'Coot A' is an electronic intelligence gathering aircraft, photographed here from a UK Royal Navy Sea Harrier during a NATO Atlantic Ocean exercise in 1985.

Electronic warfare and intelligence gathering is the role of the Antonov An-12 'Cub B'.

The first airborne early warning aircraft to enter series production was the Tupolev Tu-126 'Moss' which is primarily used overwater for the navy but probably flown by the air force.

Myasishchev M-4 'Bison' jet bombers were converted to maritime reconnaissance and in-flight refuelling tasks. *(USN)*

The latest airborne early warning and control aircraft is the Ilyushin Il-76 'Mainstay' for both tactical and naval use. It is currently in service in limited numbers but about five are thought to be produced annually.

Uruguay

Organisations: Uruguayan Naval Aviation.

Organisational structure: The navy has control of all maritime patrol, anti-submarine warfare and search & rescue operations as well as operating a small transport fleet.

Command structure: Details not available.

Air-capable ships: None.

Shore bases: Laguna, Punta del Este.

Embarked aircraft: None.

Shore-based aircraft: Beech Super King Air 200T (1); Beech TC-45J (5); Beech T-34A/B Mentors (6) (1); Beech T-34C Turbo-Mentor (3); Bell 222 (3); Bell 47G (2); CASA C 212 Aviocar (1); Grumman S-2A/G Tracker (6); North American T-6 Texan (4); North American T-288 Fennec (18); Piper PA-18 Super Cub (2); Sikorsky SH-34J (2).

Units: No details available.

Recent operations: None reported.

Typical deployments: Patrol of the River Plate estuary and coastal/river surveillance.

Weapon systems: Details cannot be confirmed but it is understood that the Tracker aircraft have been fitted for lightweight torpedoes and depth bombs, as well as carrying 127 mm rockets for anti-shipping operations; the Texan/Fennec force could be armed with rockets and machine guns.

Personnel: 4500 (total).

Remarks: Uruguayan military forces are usually starved of funds for new equipment and modernisation programmes, although the threat posed by neighbouring nations is probably not sufficient to warrant greater spending.

Venezuela

Organisations: Venezuelan Naval Air Service; Venezuelan Coast Guard.

Organisational structure: The navy has equal status to the air force, reporting to the National Council of Security and Defence, the Chief of the Joint Chiefs of Staff and the individual service commanders. The naval air service administers three operational units. The coast guard force was established in August 1982 to patrol the country's 200mm exclusive economic zone.

Command structure: navy: there is thought to be a flag officer with overall operational control of maritime air operations, including training and transport; the squadron is the basic operational unit but no details of the structure are available; coast guard: subordinate to the navy on all matters.

Air-capable ships: 6 *Lupo* class frigates, 4 *Capana* class tank landing craft (platform only); coast guard: 2 *Almirante Clemente* class frigates.

Shore bases: naval: Caracas (La Carlota); coast guard: Puerto Cabello.

Embarked aircraft: Agusta AB 212ASW (10).

Shore-based aircraft: Agusta ASH-3D Sea King (4); Beech King Air 90 (1); Bell 47J (2); CASA C 212 Aviocar (7); Cessna 310R (2); Cessna 402C (1); de Havilland Canada Dash-7 (1); Grumman S-2E Tracker (6); MBB BO 105C (2); coast guard: Cessna 210 (4); Grumman Hu-16 Albatross (4).

Units: naval: AS01 (S-2E); TR02 (transports); MP03 (AB 212ASW/Sea King); coast guard: no details available.

Recent operations: None reported.

Typical deployments: AB 212ASW helicopters are embarked in the *Lupo* frigates for routine operations in the Atlantic and Caribbean, including multinational exercises.

Weapon systems: The Italian-built helicopters are thought to be armed with the Whitehead Motofides A-244S lightweight torpedo and depth bombs; the Trackers carry the Mk 44/Mk 46 lightweight torpedo and have provision for 127 mm rockets for anti-shipping duties; the weapons fit for the Aviocar probably includes depth bombs and light cannon/ machine guns.

Personnel: navy: 7 500 (total); coast guard: not available.

Remarks: Venezuela is seeking additional maritime patrol and security helicopters to supplement the fixed-wing operations of the air force, according to reports in October 1987. Other reports suggest that there are already 12 AB 212ASWs in service.

Vietnam

Organisations: People's Liberation Army.

Organisational structure: Unknown but said to be 'revolutionary'.

Command structure: Unknown.

Air-capable ships: None.

Shore bases: Cam Ranh Bay; Chu Lai; Da Nagng; Dong Ha; Hai Pong; Hue; Quang Tri.

Embarked aircraft: None.

Shore-based aircraft: Mil Mi-4 'Hound B' (10).

Units: Presumably a single helicopter 'squadron'.

Recent operations: Little maritime air activity by Vietnam has been noted but Soviet long-range maritime reconnaissance operations are thought to be extensive.

Typical deployments: Vietnam-based Soviet naval aircraft range into the Indian Ocean, South China Sea and Pacific Ocean, often joining up with air elements from Soviet Union (Pacific coast) and Indian Ocean bases. US forces concerned about Soviet build up of Cam Ranh Bay and Da Nang airfields for anti-shipping cruise missile armed aircraft deployment.

Weapon systems: Basic Soviet ASW equipment in Hound-Bs; Soviet aircraft weaponry discussed in relevant section: Soviet Union.

Personnel: Unknown.

Remarks: Little is known about the small air arm of the Vietnamese People's Liberation Army, other than most of the US-supplied equipment to the former Republic of Vietnam (South Vietnam) is redundant and out of service. Mi-8 helicopters have a low intensity ASW, surface search and SAR operational role. Since 1985, there have been growing reports of Soviet air activity from former US air bases along the coast. According to the US Department of Defense, it is significant that in February 1986, Soviet naval and naval air assets at Cam Ranh Bay participated in the first co-ordinated anti-carrier warfare exercise to be held in the South China Sea.

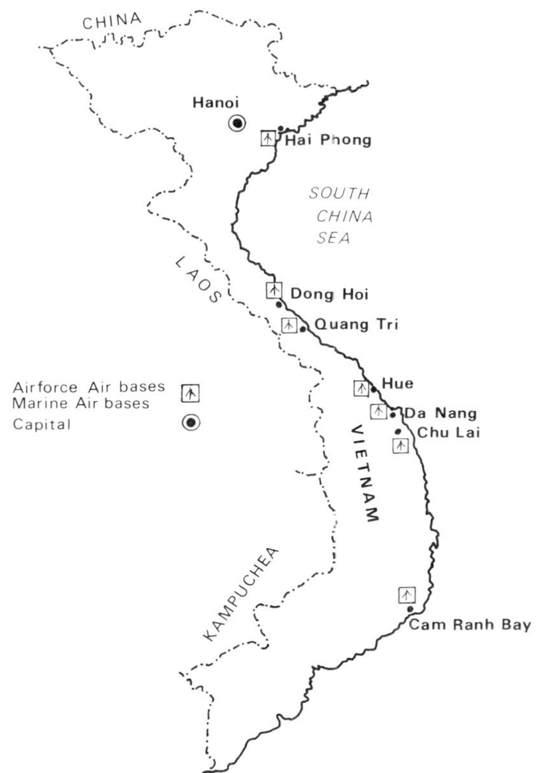

Yugoslavia

Organisations: Yugoslav Air Force; Yugoslav Navy.

Organisational structure: Political command of the armed forces vested in government ministers. Although the navy has only a small air unit, it carries out most of the overwater operations.

Command structure: Following the Soviet model, the basic operational unit of the air force is the regiment; the naval air arm uses the squadrons as its basic operational unit.

Air-capable ships: None.

Shore bases: Mostar; Niksic; Pula; Titograd.

Embarked aircraft: None.

Shore-based aircraft: Kamov Ka-25 'Hormone A' (15); Kamov Ka-28 (5); Mil Mi-8 'Hip C' (20).

Units: Identification of individual units has not been possible, although three helicopter squadrons are thought to be operational.

Recent operations: None.

Typical deployments: Coastal defence only.

Weapon systems: Mainly confirmed to older generation Soviet systems, but several appear to have been modified for interaction with Western sensors and other associated systems.

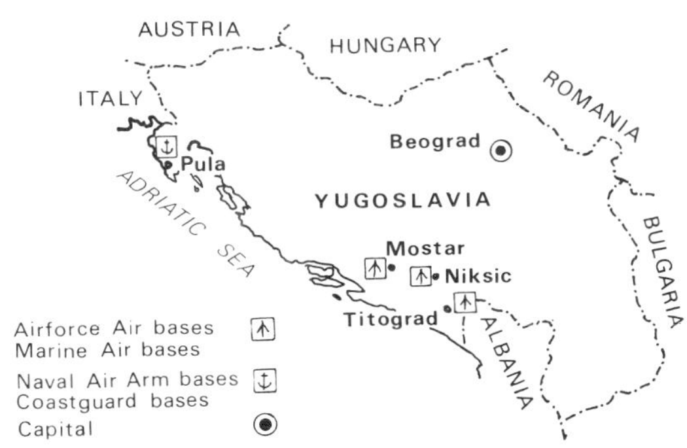

Personnel: 14 000 (total).

Remarks: The naval air arm has a primary anti-submarine role and a secondary one of marine assault, although the marine force is part of the army. It seems that there is no direct air force fixed-wing maritime air support for the Yugoslav Navy, although ground attack, fighter and transport aircraft could be presumably used in time of conflict. Yugoslavia, although it has a large amount of Soviet equipment in its inventory, is not a member of the Warsaw Pact.

Amphibious SAR and surveillance tasks are carried out by the Canadair CL-215 which also doubles as a fire fighting aircraft.

Kamov Ka-25 'Hormone A' shore-based ASW helicopter, one of 15 reportedly operational. *(Ivo Sturzenegger)*

Kamov Ka-28 'Helix A', part of a batch of five delivered in 1988. *(Ivo Sturzenegger)*

Index